Perelandra Soil-less Garden Companion

Other Books by Machaelle Wright

Behaving as if the God in All Life Mattered

Co-Creative Science:
A Revolution in Science Providing Real Solutions
for Today's Health and Environment

Dancing in the Shadows of the Moon

The Mount Shasta Mission

Flower Essences:
Reordering Our Understanding and Approach to Illness and Health

MAP:
The Co-Creative White Brotherhood Medical Assistance Program

Perelandra Microbial Balancing Program Manual

Perelandra Garden Workbook:
A Complete Guide to Gardening with Nature Intelligences

Perelandra Garden Workbook II:
Co-Creative Energy Processes for Gardening, Agriculture and Life

Perelandra Soil-less Garden Companion

Machaelle Small Wright

PERELANDRA, LTD.
CENTER FOR NATURE RESEARCH
JEFFERSONTON · VIRGINIA

Contacting Us

Perelandra, Ltd.
P.O. Box 3603
Warrenton, VA 20188

Web Site:
www.perelandra-ltd.com
E-mail:
email@perelandra-ltd.com

Question Hot Line:
1-540-937-3679

Order Lines:
U.S. & Canada
1-800-960-8806
Overseas & Mexico
1-540-937-2153

Fax:
1-540-937-3360

Perelandra Soil-less Garden Companion
Machaelle Small Wright

FIRST PRINTING 2008
E-Book Edition 2007

This book is manufactured in the United States of America.
Designed by Machaelle Wright.
Cover design & line drawing by James F. Brisson, Williamsville, VT 05362.
Copyediting by Elizabeth McHale, Williamsburg, MA 01096.

Proofreading and GROMs extraordinaire:
Beth Shelton, Jeannette Edwards and Joyce McVey.
Formatting, typesetting, and computer wizardry
by Machaelle Small Wright.
This book was formatted, laid out and
produced using QuarkXPress software.
Printed on recycled paper.
Published by Perelandra, Ltd., Warrenton, VA 20188

Library of Congress Control Number: 2007939041
Wright, Machaelle Small
Perelandra Soil-less Garden Companion
ISBN: 0-927978-70-9

2 4 6 8 9 7 5 3 1

Contents

Perelandra
Soil-less Garden
Companion

Introduction

I have written the *Perelandra Soil-less Garden Companion* because, quite frankly, those who have been working for years with the Perelandra soil-less garden processes in conjunction with their projects threatened to hit me if I didn't write something. And, after sitting down and looking at what they were complaining about, I had to agree — much to my dismay — that they were absolutely right to complain and threaten me. This *Companion* is going to make life a lot easier for everyone.

If you are new to the idea of soil-less gardens (SLGs), let me explain what they are. Most likely you think I'm referring to hydroponic gardening. The soil-less gardens I'm referring to are projects and goals such as a business, job, home, classroom, education program, writing a book, delivering a speech . . . For more information on soil-less gardens and why we benefit from working in partnership with nature with this type of garden, read *Co-Creative Science* and watch the DVD/video *Working with Nature in Soil-less Gardens.*

And this brings me to an important point about the *Companion.* I call it "companion" for a reason. It's not meant to replace the key soil-less garden information I've previously put out. Rather, it is to be used along with that information.

In the *Companion*, I centralize important information, include some new SLG processes and update the Soil-less Garden Troubleshooting Process. As I put together the information in the *Companion*, I took the opportunity to streamline and simplify the SLG Troubleshooting Process and its steps. Along with this, I also introduce a Troubleshooting Process Chart that is specifically designed for the SLG

Troubleshooting Process. You "old timers" are going to love the new simplified balancing and stabilizing steps for each of the energy processes.

> If you have steps from other Perelandra SLG material with differing instructions, use the information I've given you in the *Companion.* The *Companion* contains the latest and most up-to-date SLG information. Also, the new SLG Troubleshooting Process Chart and the streamlined energy processes steps are to be used with soil-less gardens only. Do not use them for *Workbook II* environmental troubleshooting or for the Microbial Balancing Program troubleshooting.

To work with a soil-less garden and the SLG Troubleshooting Process, you need the following:

- *Working with Nature in Soil-less Gardens* (DVD/Video 3).
Contains critical SLG information and includes instructions for doing kinesiology.

- *Perelandra Soil-less Garden Companion.*
Includes the updated information and steps to be used for the SLG Troubleshooting Process. Use in conjunction with DVD 3 and the SLG Chart.

- Copies of the SLG Troubleshooting Process Chart.
You may download a free copy of this chart from our web site. I suggest printing 50 copies! Or you may xerox a blank chart from the back of this book.

- ETS Plus for Humans, ETS Plus for Soil-less Gardens and ETS Plus for Soil.
Brochures are included with each ETS Plus and are available on our web site.

- The Perelandra Essences: Rose Essences and Rose II Essences.
Perelandra Rose Essences and Rose II Essences are *the two required sets* for working with soil-less gardens. However, working with all five sets of Perelandra Essences provides more comprehensive coverage. NOTE: I am not referring to the Perelandra Essences you test for yourself. The Rose Essences and Rose II Essences only refer to the *minimum* needs for a soil-less garden.

- Perelandra Nature Cards.

The Cards are especially helpful when we need to become "unstuck" and don't know what to do next.

- Pen/pencil, paper, spoon, paper towels, a watch or clock with a second hand.

OPTIONAL

- *Co-Creative Science* (book).

Lays a good foundation for understanding nature, form and why SLGs work.

- *Perelandra Garden Workbook II* (book).

Provides background information on the energy processes.

A NOTE TO THOSE WITH ALREADY-ACTIVATED SLGS

Consider the information in the *Companion* as an opportunity to upgrade your active SLG(s). Don't think you have to throw it out and start over. Look at your current DDP and Project Coning. If either need to be changed, just make the needed changes and put them through the Starting Process. If you only need to improve your DDP, but your Project Coning remains the same, you will still need to put both through the Starting Process. This allows both of these pieces to connect and weave together in the necessary ways. Once you complete the Starting Process, then continue with your SLG using the new process steps and the new information. You'll find that the ebb and flow of your SLG will move better and, I suspect, some bumps you might have been encountering will now clear up.

Chapter 1

Writing Soil-less Garden DDPs

To start a soil-less garden is to open yourself up to quite an adventure. Don't forget that your coning partners include nature and the White Brotherhood. (See Chapter 2.) Trust me: There's not going to be anything "average" about your soil-less garden with team members like this. So if you're not up for an adventure, if you don't want to experience anything new or different, if you want to maintain complete control over your project, don't open a soil-less garden. If you do open one, remember to expect the unexpected. This will be your new "average."

In order to start a soil-less garden, you need to choose a project or goal and then succinctly and precisely describe its definition, direction and purpose (DDP). You can't just stand there and say to your coning partners, "I wanna start a business." If that's all the information you're going to supply your partners, you could end up with quite a surprise. You might find yourself starting a sausage-making factory. There's nothing wrong with sausage making, but it may not be the kind of business you had in mind. You're the only one who can provide the definition, direction and purpose for a soil-less garden. Your SLG team members will not do that for you. If you "wanna start a business," the DDP for your business might look more like this:

> I want to start a daycare center for low-income children, ages eighteen months through four years. At this center, I want to provide the children with lessons about nature and give them a new understanding of nature in their lives.

Notice what I did not write in that DDP. I did not write where I wanted the daycare center located. This means I'm leaving it wide open for my partners to provide all that is necessary for locating the center where it will flourish the best. I also did not put a limit on the number of children the center can accommodate. Again, I'm leaving that part of the planning wide open. This is because, after thinking about it, I realized I'm willing to devote my working life to this center and, if it grows into something big, I'm willing to accept that development knowing that my partners will supply all that is required to maintain and operate a big daycare center. I also did not list all the business elements needed for my daycare center. That's because everything that is needed for a daycare center is automatically included/implied under the name "daycare center." You're not going to activate a soil-less garden for a daycare center and discover that you're receiving equipment and ideas that are needed for starting that sausage factory.

But there's something else to consider here. By not micro-defining what I want for my daycare center, I'm allowing my partners to provide me with the information and insight on what a daycare center *can be* and to provide me with the matter, means and action to put together a new kind of daycare center. I may end up with a center that looks and functions completely different from what we would consider "traditional" for daycare centers. So by not micro-defining "daycare center," I leave open the possibility for something new and better to be created.

Now, notice what I did write. I defined the business by describing it as a daycare center. I gave it direction by writing that it is a center for children (and not senior citizens, handicapped, etc.), giving the age range of the children and specifically listing that the children are to be from low-income families. I then included additional purpose to this daycare center by writing, "At this center, I want to provide the children with lessons about nature and give them a new understanding of nature in their lives." But I didn't further define "lessons" and "new understanding" because I want to leave it open for my partners to give me insight about the particulars of the lessons and what a "new understanding" of nature would be for the children in my daycare center.

Let's say I want to start this daycare center, but I need for it to be located in my home. I also want to expand the center to include any children from my area, not just low-income children. However, since I will be the only adult present during daycare hours, I need to limit the number of children to four. My DDP might now read:

> I want to start a daycare center for children in my area, ages eighteen months through four years. I want to set up my home to accommodate a maximum of four children because I will be the only adult present. At this center, I want to provide the children with lessons about nature and give them a new understanding of nature in their lives.

With this DDP, I've added more specific definition, direction and purpose but, at the same time, I've also added more limitations. That is because I'm being realistic about what I really want and how it will fit into my life. To add the refinements to the original DDP, I'm functioning in my role as the person responsible for providing definition, direction and purpose for my project.

When I listen to "soil-less gardeners" talking about their DDPs, I find that the most common mistake they make is writing over-wordy, over-long DDPs. In their zeal to be precise, they put together DDPs that go on for pages. (I've actually seen DDPs that go on for pages.) For some, this is just a mistake in their understanding of how the words "succinct" and "precise" apply to a DDP. They translate those two words to mean "all-inclusive" and "exact in details," which then gets translated into pages and pages describing the all-inclusive, detailed DDP. For others, a lengthy DDP is their way of maintaining full control over their project and inserting the needed manipulative devices for "controlling" their coning partners. Then there are the people who just have a personal aversion to the notion of simplicity and feel a *simple* DDP won't give their partners enough information or won't work. With these overly wordy DDPs, you don't have a real soil-less garden. You have a fake soil-less garden that only appears to be co-creative.

The DDP that started me off into co-creative science and this thing I call "Perelandra" back in 1976, was as follows:

> I want to learn about devas and nature spirits, and I want to learn about them within the context of a garden.

All that has developed both within me and at Perelandra is the result of this simple DDP, and I've not needed to change it in over thirty years. On the surface, this seems like an exceptionally simple DDP that could not possibly cover everything that has developed here.

And that brings me to the next important thing about DDP wording. Every word in my DDP is connected to a universe of meaning that changes the nuance of the word as I grow and am more able to consider a larger definition. Look at each word separately:

I: Well, books have been written on the question, "Who am I?"

Want: What does it mean "to want"? What did it mean for me thirty years ago and what does it mean for me now?

To Learn: (I never have understood prepositions, so I'm hooking "to" with "learn" here to hide my ignorance in this area.) What does it mean to learn, and what does it mean for *me* to learn?

About: Another one of those annoying prepositions. I feel certain there is some grammar teacher somewhere who could ramble on for hours about the meaning of prepositions. If you care about this, you'll need to search for said teacher.

Devas and Nature Spirits: What I've learned about these two dynamics in nature intelligence is way beyond what I knew in 1976, when I included them in the DDP. At that time, I couldn't imagine everything they encompassed. The word "and" clearly means I want to learn about both dynamics, not just one of them.

And I Want to Learn about Them: A bridge of words that connect me to . . . *(drum roll)*

Within the Context of a Garden: And this is the kicker. When I activated my DDP in 1976, I had no idea what my partners understood about the word "garden." I didn't even nail this DDP to my existing vegetable garden. I said "in the context of a garden." I *assumed* that I was referring to my vegetable garden. And this is how a simple DDP can lead to an adventure beyond your imagination.

Your coning partners are nature and the White Brotherhood. They don't use our dictionaries and encyclopedias because those books are too confining and limiting in their descriptions and definitions. Hell, they don't even Google. No need to. I went along for about fifteen years with a traditional and somewhat limited understanding of the word "garden." And out of respect for me and my personal range of understanding, all of my early work with nature was focused in my vegetable garden. Within that context, my partners gave me a solid understanding about nature and our partnership that became my foundation for the future. My limited understanding of "garden" wasn't holding us back. But as I did the early work, something began to happen that caused me to consider the possibility of a wider definition of "garden." My work with nature gradually began to encompass areas outside the vegetable garden, such as in human health. This raised questions. Finally, in 1991, it occurred to me that my definition of "garden" and my partners' definition may be a little different. So I sat down "with them" in a four-point coning session and asked the "simple" question: "What's your definition of a garden?" And this is what they said.

THE PERELANDRA CONING: *From nature's perspective, a garden is any environment that is initiated by humans, given its purpose and direction by humans and maintained with the help of humans. For nature to consider something to be a garden, we must see humans actively involved in all three of these areas. It is the human who calls for a garden to exist. Once the call is made, nature responds accordingly to support that defined call because a garden exists through the use of form.*

Humans tend to look at gardens as an expression of nature. Nature looks at gardens as an expression of humans. They are initiated, defined and maintained by humans. When humans dominate all aspects and elements of the life of the garden, we consider this environment to be human dominant. We consider an environment to be "nature friendly" when humans understand that the elements used to create gardens are form and operate best under the laws of nature, and when humans have the best intentions of trying to cooperate with what they understand these laws to be. When humans understand that nature is a full partner in the design and operation of that environment — and act on this knowledge — we consider the environment to be actively moving toward a balance between involution (nature) and evolution (human).

As a result, such an environment supports and adds to the overall health and balance of all it comprises and the larger whole. It also functions within the prevailing laws of nature (the laws of form) that govern all form on the planet and in its universe. In short, when a garden operates in balance between involution and evolution, it is in step with the overall operating dynamics of the whole. The various parts that comprise a garden operate optimally, and the garden as a whole operates optimally.

Nature does not consider the cultivation of a plot of land as the criteria for a garden. Nature considers a garden to exist wherever humans define, initiate and interact with form to create a specialized environment. This is the underlying intent of a garden and the reason behind the development of specialized environments such as vegetable gardens. Nature applies the word "garden" to any environment that meets these criteria. It does not have to be growing in soil. It only needs to be an environment that is defined, initiated and appropriately maintained by humans.

This is what nature means when it uses the word "garden." The laws and principles that nature applies in the co-creative vegetable garden are equally applicable to any garden, whether it is growing in soil or otherwise. . . . The principles and processes apply across the board because all gardens are operating with the same dynamics — only the specific form elements that make up each garden have changed.

Well, I sure haven't found this definition of "garden" in any dictionary. Who knew? Except for my coning partners. Obviously they knew. As a result of getting that definition, I was able to open up the research I was doing at Perelandra in a number of different directions — all within the context of a garden. With this new understanding of "garden," the concept of "soil-less gardens" was born/created. And my understanding of my original, *simple* DDP expanded.

Now I come to another important point about DDPs. I'm often asked to define "definition," "direction" and "purpose." I've annoyed many people over the years by telling them to open their dictionaries. Here are my thoughts on the subject: Definitions of words have a starting point for each of us that is generally related to our age and the scope of experience we've had when we first come across the word. It's like a perfect storm: the word + our age + our experience = our definition starting point. As our age and experiences "grow," our definitions can grow, as well. And our project can grow to reflect our evolution with a specific word used in a DDP or the whole DDP. We like to talk about evolution being never-ending. Well, if this is the case, *then nothing can be stagnant.* Everything contains within it the opportunity for growth and change — including the definitions of words. A changing definition can inspire changes in us — in how we think, how we act, how we live.

I find with many soil-less gardeners there is much impatience about word "definitions." They need to know what something means, they want to know everything about it, and they need to know *NOW*. They think. They consider. They research. They ruminate. . . . They scramble so much in an effort to "know everything" about a word that they have totally lost sight of what they already know about that word. They leave no room for organic and natural personal growth in this area. They believe that if they just keep heaping information on top of their heads, this alone will provide them with "full and perfect" understanding. Let me break this to you gently: There is no full and perfect understanding — of words or anything else. There's only good, working understanding that is relative to our life at that moment and that then serves as a foundation for future and new understanding.

The foundation I am speaking of deserves more of our respect than many seem willing to give. It is a living, expanding phenomenon that reflects where we are, what we know and what we are ready to know. It provides the springboard for all integrated and meaningful growth. It has a dynamic all its own, and we cannot intellectually control the range and scope of that foundation. Trying to take charge

intellectually just adds confusion and distraction to the mix. What dictates, what controls, our foundation's range at any given point in time is the sum of all the parts of our being and our life.

When considering a DDP for a project, start with what you already know about the words "definition," "direction" and "purpose." That will be the foundation for your DDP. And over time your foundation and what you understand about these words is going to grow. I guarantee you that ten years down the road your DDPs are going to be constructed differently from what you put together today. And you never would have gotten to that point ten years down the road had you not jumped in and started putting together a DDP now. *Or,* like me, you'll find in ten years that your first DDP was brilliant beyond words and just what was needed, even without you knowing it at the time. In short, writing a DDP is its own adventure that will lead you to extraordinary heights and growth.

So you've come to this point and you realize you still have no clue what "definition," "direction" and "purpose" mean, even after you open a dictionary, take a few doses of ETS Plus for Humans, do a Calibration Process for each of the words and consult with your MAP team. There is no foundation for you to build on. And you are now frustrated and depressed. Well, hold on! You're in luck. You've just "chosen" your first soil-less garden. Your project is to discover what these words mean and how those definitions apply to you and soil-less gardens. I've just given you the DDP:

> I want to discover what "definition," "direction" and "purpose" mean and how these definitions apply to me and soil-less gardens.

Actually you may start this soil-less garden *before* you open a dictionary, take a few doses of ETS Plus for Humans, do a Calibration Process for each of the words and consult with your MAP team. If these things are needed for your journey into the world behind those three words, you'll get insight about it from your team.

AMENDING A DDP

You may amend a DDP after it has been activated. Let's say you gave a DDP your best effort and, at some point down the road, you realize you are now dealing with the elements of that weird sausage factory "thingy." This is telling you that you need to straighten out or clarify your DDP. The implications of your DDP are being accurately reflected back to you by your coning team. To amend a DDP, do the following:

- Open the Project Coning. (See Chapter 3.)
- Based on what you observed happening in the project, amend the DDP. Add a critical word, phrase or sentence. Or take out a critical word, phrase or sentence. (You may want to remove any reference to "sausage factory!")
- Consider this a new DDP and activate it using the Starting Process. (See Chapter 3.)

Then continue with your project from where you left off prior to amending the DDP. There's no need to throw out all the work you've done up to that point. However, the new DDP may inspire you to look at that body of work and toss out those things that no longer relate to this DDP. Most likely, these will be the things that caused you concern to begin with, like someone in the business ordering 120 barrels of sausage casings.

DDP EXAMPLES

The following are examples of DDPs just to give you an idea of the wording to start with and the wide range DDPs and soil-less gardens can cover.

THE PERELANDRA GARDEN DDP:
I know this is not a soil-less garden, but I thought seeing how I constructed the Perelandra garden DDP might be helpful to you.

1. To demonstrate co-creative principles and co-creative science within the context of Machaelle Wright's garden.

2. To produce essences, MBP Balancing Solutions and ETS Plus tinctures.
3. To balance, stabilize and support the garden research and development, and Perelandra's outreach activity.

Examples of DDPs from Soil-less Gardeners

FOR A PERSONAL HEALTH-RELATED PROJECT:

I would like to understand what I need to do personally to establish and maintain an optimal level of physical health. I also want to implement and maintain these elements in my life.

A HEALTH AND WEIGHT-LOSS DDP:

1. I want a better understanding about a diet that is appropriate for me.
2. I also want a better understanding about the physical agility, strength and flexibility that is appropriate for me at my age, taking into account my responsibilities and activities.
3. I want to achieve and maintain this appropriate physical agility, strength, flexibility and weight range.

A FINANCE DEPARTMENT DDP WITHIN A BUSINESS:

Establish and activate cash flow in the following areas (as listed on our weekly finance report):

To meet the coming week's operating expenses, bills due and obligations
To budget income in preparation for planned future obligations
To cover additional payments the company would like to make
To earn additional income for building and maintaining our bank accounts
To provide the information, understanding and input needed to stabilize, strengthen and build ________'s (the company) overall financial picture.

A WEB SITE DDP:

To provide easy access and introduce ________ (the company) to as many people as possible, in the following ways:

Inform and educate
Excite and inspire
Serve as an interactive and multifaceted education and information vehicle
Facilitate visitors through layout, design, organization, humor, text and tools

Reflect the spirit of the company through dignified, calm and non-aggressive means

Generate sales

SHORT-TERM DDP FOR FENCING A WILDFLOWER MEADOW:

1. Construct a four-board panel fence around our meadow perimeter
2. That provides our dogs with a border that they will not go beyond
3. Painted white
4. With one metal gate on the west border for entering the back field
5. Installed and painted professionally, quickly and neatly
6. By contractor/workmen easy to work with, reliable, interested in and capable of doing a quality job
7. With contractor being capable of doing the job without causing unnecessary damage to the meadow
8. For a fair and reasonable price with the matter, means and action for me to easily meet that price without draining my personal bank accounts
9. Installed by early fall 2003.

Results: This DDP was activated in early September 2003 and deactivated on November 11, 2003. All of the points of the DDP were met except for a couple of small places in the wet areas of the field where vehicle ruts were left.

A DDP FOR REFINANCING A HOME:

An easy, efficient, friendly refinancing of our home and property

Interest rate: lower than 6%

Closing costs: less than $2000

Prepaid items: Less than $1750

Conventional loan. 30-year fixed rate. No points.

Closing date: May 2005

Results: All of the above DDP points were either exactly met or bettered. For example, the interest rate was 5½% and the closing costs were less than expected.

APPRAISAL DDP FOR THE ABOVE REFINANCING:

1. I want an appraiser who understands, sees and appreciates the value of our home and 12 acres,

2. Who appraises/values our home and land high enough to cover the full refinancing package with the bank, and
3. Who truly likes our home.

Results: The appraiser really did like their home and location. He noticed and appreciated many of the fine touches in the home. He was pleasant and seemed interested in what they had to say about the home. He explained how he was going to put together the appraisal for this "unique" home. When he left, they felt they had the best chance for getting a fair appraisal.

Update: The appraisal was perfect, and the refinancing process went smoothly.

A DDP FOR A JOB:

I would like my proposal work and output to be of the highest quality.

I would like to determine with greater certainty whether or not my current job is part of a path I want to be on.

I would like my work not to be harmful to the ecology or environment (local or planetary) and to be helpful to the environment (local or planetary) where possible.

I would like to have a good job doing something I enjoy and feel good about with a good salary, health insurance and vacation benefits.

Chapter 2

SLG Four-Point Conings

THE THING THAT DIFFERENTIATES a regular project from a soil-less garden is the four-point coning. In a soil-less garden, it is set up to ensure perfect balance between the involution dynamic (nature) and the evolution dynamic (the White Brotherhood and you). The evolution dynamic supplies the definition, direction and purpose to any thing or action. The involution dynamic (nature) supplies the matter, means and action for achieving evolution's definition, direction and purpose. Another way to say this is that nature supplies the *organization* (how all of the elements fit together), *order* (the sequence in which the different elements play out), and *life vitality* (the activity among the various elements) for achieving evolution's *definition, direction* and *purpose.* The human soul is the force behind the evolution dynamic. Nature is the force behind involution.

Technically speaking, a coning is a balanced vortex of conscious energy. The simplest way to explain a coning is to say that it is a conference call. With a coning, we are working with more than one intelligence simultaneously.

A coning is needed for multilevel processes because of the greater stability, clarity and balance it offers. With multilevel processes, we are working with many different facets and levels of intelligences at one time. Consequently, it is far better to work in an organized team comprised of all those involved in the area we are focusing on.

A coning has a high degree of protection built into it. Because of the larger scope of the work, it is important to define exactly who and what are involved in that work. All others are excluded by the mere fact that they have not been activated in the coning. In essence, a coning creates not only the team but also the "room" in which the team is meeting. The coning is created and activated by us — the human team member. Only those with whom we seek connection will be included. Members will not "slide" in and out of a coning on their own. This adds to the exceptional degree of protection contained within the coning.

I have said that a coning also offers balance and clarity. A friend of mine who began to use a four-point coning whenever she did multilevel work was quite impressed at the difference the coning made in both the quality of her work and the process she went through during her work. She said that when she began to do her work while in a coning, she felt as if she was sitting in a beautiful, sunlit room with the windows open and a lovely warm breeze softly blowing in. Before she learned to use a four-point coning, she would just "connect" with various "helpers." She said that, in comparison, it was as if she was sitting in a room with no windows and struggling to get a full breath of air. What she was describing was the contrast she felt with the balance and clarity the four-point coning gave her.

Any combination of team members can be activated for the purpose of simultaneous input. But this does not constitute a coning. A true four-point coning has balance built into it. By this I mean a balance between nature and humans. In order for us to experience anything well, we must perceive it in a balanced state; that is, it must have an equal reflection of the human soul or spirit dynamic (evolution) combined with an equal reflection of the form/nature dynamic (involution). I see this balance in the shape of a V *(Fig. A)*.

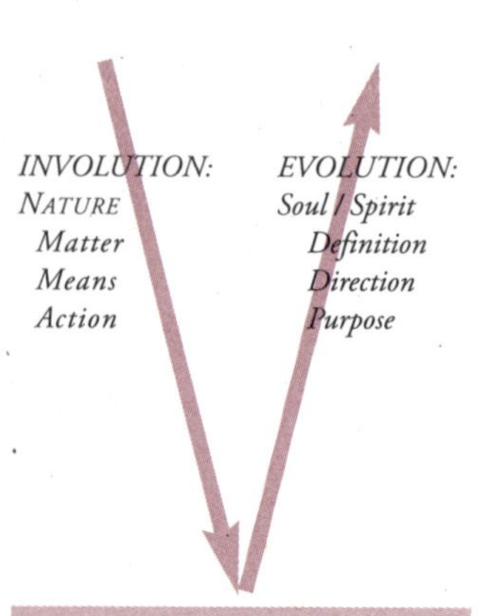

Fig. A: The Involution/ Evolution coning dynamic

For anything to function well in form, it must have within it a balance between the involution dynamic and the evolution dynamic, or nature and soul. The extent to which we achieve balance between these two dynamics depends on our willingness to allow nature to partner with us. To have soul effectively, efficiently and fully activated into form and action, we must have a balance between involution and evolution.

FOUR-POINT CONINGS

A coning is set up for the purpose of activating a team for specific work. It is therefore important for the successful completion of the work that a balance be maintained between the involution dynamic and the evolution dynamic within the coning itself. We do this by setting up a basic four-point coning that lays a balanced foundation between the involution and evolution dynamics.

As a foundation, the four-point coning maintains the necessary balance between (a) involution/nature, through the connection with the devic and nature spirit levels; and (b) evolution, through the connection with the appropriate members of the White Brotherhood and the higher self of the person(s) working in the coning *(Fig. B)*. We humans supply the evolutionary dynamic only. To operate in balance, it is important to understand that we do not supply the involutionary dynamic.

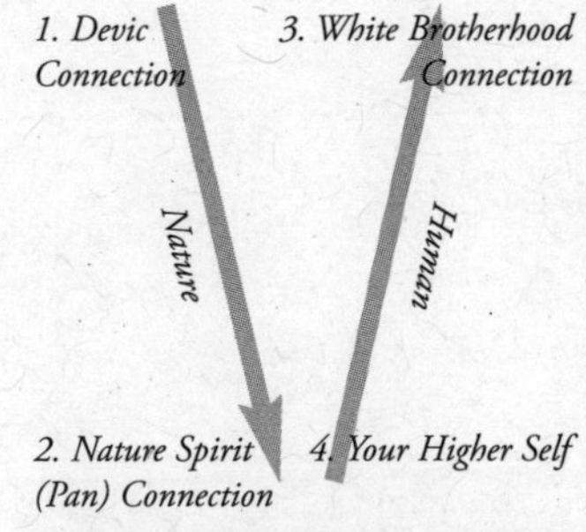

Fig. B: The balanced four-point coning

SOIL-LESS GARDEN CONINGS

A soil-less garden is for those who wish to work with nature and the White Brotherhood in business, education, the arts, the home, research, their job, personal and professional projects and goals . . . all those gardens in life that are not rooted in soil. A four-point coning may be set up for any project or situation in which we would like good, balanced input and assistance. The soil-less garden coning is an involution/evolution balanced, multilevel meeting that has been activated or opened by an individual for the purpose of exchanging information relative to a specific goal or project. People set up SLG conings for assistance in functioning in their occupations. Others set up SLG conings for starting and running their businesses and during all business meetings. With an SLG coning, a meeting runs far more efficiently and decisions are more easily reached. The people who are participating in an SLG coning work with the White Brotherhood to create and hold the appropriate vision and to define and give purpose to all that needs to be addressed within a specific project. The people also work with nature to come up with the materials, solutions and courses of action that best supply the matter, means, action and organization for what they wish to accomplish.

Anyone interested in functioning in harmony with the current transition from the Piscean dynamic to the Aquarian dynamic should include the appropriate members of the White Brotherhood in any coning they activate — and this includes an SLG coning. The White Brotherhood holds the major patterning and rhythms being utilized for this transition. They are part of a balanced coning because they support and assist by ensuring that any work conducted from the coning maintains its forward motion and its connection to the new Aquarian dynamic. When opening an SLG coning, we ask to be linked with the appropriate members of the White Brotherhood working with ________. (Define the project and the area of focus within the project that you wish to address.) The appropriate members will automatically be activated in the coning.

In all conings, we include our higher self to ensure that the work done is compatible with our higher direction and purpose. This input is given automatically by the fact that our higher self is linked to the coning.

The devic connection of the coning ensures that the work being done maintains an overall integrity with nature's design in the area in which we are working. The devic level also creates or designs appropriate courses of action and solutions to specific situations or problems. Each coning has in it a specific deva.

For some situations, you may need more than one deva. A four-point coning only refers to the divisions of the coning. It does not mean only four members can participate in that coning. In some cases, you may need to identify situations within larger contexts and include several devas as a result. For example, if one department in your business is putting together a brochure, you would include in the coning:

- the deva of your business,
- the deva of that specific department, and
- the deva of the project itself — i.e., the brochure.

These three devas constitute one point of the coning. And the fact that you have three devas "standing on that point" does not overload the point or give it additional "weight." Another example: Several years ago, one of my staff put together a co-creative staff team training program for us. To do this, she worked in a four-point coning. For the devic point, she included the Deva of Perelandra (to maintain the overall direction and intent of Perelandra), the Deva of the Perelandra Business (representing the business direction and intent) and the Deva of Team Development. These three devas were combined in the coning to create the appropriate team-development program for the Perelandra business staff.

The nature spirit point of the coning is connected through Pan. We do this because all nature spirit levels except Pan are regional, and rather than try to figure out which nature spirit levels are involved in the work we wish to do, we work with Pan — the one nature spirit level that is universal in dynamic and is involved in all of the nature spirit activities. Pan's connection helps us maintain integrity with natural law for all action and issues worked on in the coning and gives assurance that the nature spirit activity and input is fully represented at all times.

Steps for Opening & Closing an Activated SLG Coning

To put together and activate a new four-point coning for a soil-less garden project, see the Starting Process (Chapter 3, Step 3). The following are the steps for already-activated SLG conings.

SUPPLIES NEEDED: ETS Plus for Humans

1. State: I wish to open the SLG coning for ________ (project).

2. State: I'd like to connect with the following members:
Wait 10 seconds after each connection.

- The deva of ________ (project deva)
- Pan
- The White Brotherhood — State:
  I'd like to be connected with the appropriate members of the White Brotherhood working with ________ (project).
- Your higher self — State:
  I'd like my higher self to be connected to the coning.

When others are present and wish to participate in the coning session, include their higher selves in the coning at this point. After including your higher self, repeat the process for each person present. Don't try to connect them to the coning as a group. For each person, state:

I'd like to connect ________'s higher self to the coning.

Wait 10 seconds before turning your attention to the next person.

Wait 15 seconds after activating the full coning.

3. Take one dose (10–12 drops) of ETS Plus for Humans.* Each person whose higher self is connected to the coning also takes one dose of ETS Plus.

4. Do the work or have the meeting. Be sure you take good notes.

* Whenever you take a dose of ETS Plus for Humans, you are to take that dose orally. Do not try to shift this to yourself using the NS Application Process.

Closing the Coning

1. When the work is complete, thank your team for its assistance and close the coning. *In the following order,* ask the members to disconnect:**

- Your higher self (Include the higher self of anyone else who is included in the coning. State each person's name.)
- The appropriate members of the White Brotherhood for this SLG
- Pan
- The deva of your project and any other devas included in the coning meeting

Wait 15 seconds.

2. Each person takes one dose of ETS Plus for Humans.

** No waiting time is needed between disconnecting each of the coning members when closing the coning.

IMPORTANT POINTS ABOUT CONINGS

● CLOSING DOWN A CONING: A four-point coning closes automatically as soon as you request the disconnections. No member of the coning is going to resist a disconnection any more than they would resist a connection. You are the one in command of the creation, activation and dismantling of the coning.

● MULTIPLE MEMBERS: You may include as many members in your team as you like and maintain involution/evolution balance as long as you begin with the regular four-point coning as your foundation. However, it is best to include only those who are directly related to the work being done. Some people get a little crazy about coning activation and add forty-two members who might possibly be involved or might like to be involved. If you do, you no longer have a tight, cohesive team. You now have a mob coning instead.

> I urge you to resist activating a mob coning for two reasons:
> - It is totally unnecessary and does not strengthen or enhance a coning in any way.
> - You, as the initiator and activator of the coning, will feel a substantial physical drain as you try to hold a coning with an unnecessary number of members present throughout the time it takes for the process or work.

A coning, although invisible, is a physical energy reality that will interact with you. A mob coning would require the same amount of energy as trying to hold a deep conversation with six people in the middle of the New York Stock Exchange on Friday afternoon. You are going to feel exhausted from this kind of overload. So, the key to "good coning building" is to be precise about who you request to be on the team. Each member should have an integral role in the work to be done.

There's another point. No other person's higher self should be added into a coning without his or her *conscious permission.* To do so without conscious permission from that person is just plain rude and ethically wrong. Also, *they must be physically present during the coning meeting.* Should someone need to leave the meeting before the coning is closed down, simply focus your attention on the coning and request that this person's higher self be disconnected from the coning. Before leaving, the person should take one dose of ETS Plus for Humans.

BECAUSE WORKING IN A CONING IS A PHYSICAL PHENOMENON FOR US, WE NEED TO ADDRESS SOME BASIC PHYSICAL NEEDS:

● CONING HOURS: When starting a soil-less garden, we should not work in its coning for more than one hour at a time. This is especially true in the beginning

when we are getting used to working with and in that coning. Your body needs time to adjust to this new dynamic. If you need to do more than can be accomplished within the one-hour time frame, close down the coning and activate a new coning session later that day or the next day. Once your body has adjusted to this work and is comfortable, you may work longer in a coning. Let your body and its comfort dictate the length of time you should work in a coning.

● PROTEIN DRAIN: Activating, working with and dismantling a coning require sharp focus on our part. As a result, we may experience a protein drain. You will know you are experiencing such a drain if you come out of the coning and are suddenly attacked with a galloping case of the munchies. Often, we translate a protein drain into a desire for sweets. I have nothing against a great big piece of good chocolate from time to time. But in this case, you need protein to compensate for the drain. Then eat your chocolate.

HINT: Have a bag of nuts with you while you are in a coning and occasionally eat a few. This will avoid or minimize protein drain while you are in the coning.

Also, I find that while in a coning, I do not perceive a protein drain or any hunger. It doesn't register until after I close the coning. So I suggest that, if you are not hungry during the coning, don't let that fool you. Still have your supply of nuts handy and eat a few throughout the session.

● TAKING BREAKS WHILE IN AN OPEN CONING: If you need to take a break during a coning session, *do not close the coning* and then re-open it. Just tell your team that you need a break for fifteen minutes or a half-hour or whatever (don't make it a vacation). They will automatically shift the coning into what I call the "at-rest" position. You may feel the intensity of the coning back off a bit when this shift occurs. It will take about five seconds. Then have your break. When you are ready to resume, announce that your break is over and that you would like to activate the coning again, and the coning will automatically shift back into its previous active connection and intensity.

● JUNK CONING SESSIONS: Don't get into the habit of activating conings indiscriminately. They are to be used for interlevel work and not for prediction-type information. Your coning members are functioning in a team dynamic with you and do not wish to be asked to function in areas where your common sense should be the dominant factor. Opening a coning to ask if you should go on a specific vacation or if it would be good for you to buy that sweater for your father is not appropriate. As far as your team is concerned, such decisions are best left to your common sense and intelligence.

● KEEPING IT CURRENT: I also suggest that you limit the questions to issues that pertain to the present. Information about the future is reliable only if all issues and elements involved remain exactly the same as they were when you asked your questions. In a changing world populated by over six billion people with active free will, this never happens. By the time you get to some future point in time, all relevant issues and elements will most likely have changed perspective and position. Therefore, your answers about what to do will also have changed. In short, it is a waste of time and a crapshoot to try to see into the future — from within a coning or otherwise.

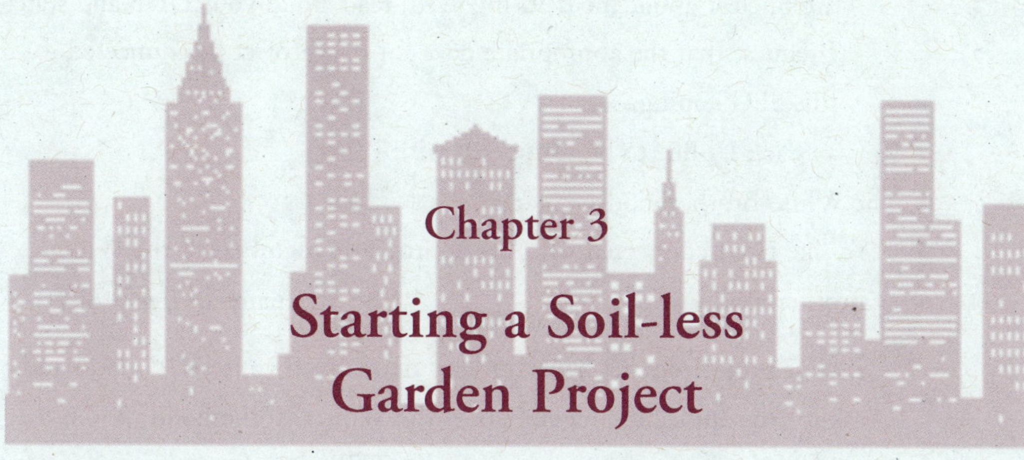

Chapter 3

Starting a Soil-less Garden Project

SUPPLIES NEEDED: ETS Plus for Humans, ETS Plus for Soil-less Gardens, ETS Plus for Soil, pen/pencil, paper, spoon, paper towels, a watch or clock with a second hand.

THE STARTING PROCESS STEPS

1. CHOOSE A PROJECT to work with or a goal you wish to achieve.

2. WRITE A DDP. That is, put in writing your project and succinctly describe its definition, direction and purpose.

3. SET UP A FOUR-POINT CONING, thus establishing your SLG coning team.

   A. State: I wish to open an SLG four-point coning.

   B. Link with each appropriate member of the coning individually, waiting 10 seconds after each connection.

      - The specific deva involved (the deva of your project)
        If your project has a name or title, this would be the Deva of ________ (title). State:
        I would like to be connected with the Deva of ________.
        Example: We have the Deva of Perelandra in our SLG coning. If you

are unclear about the deva involved, read aloud your DDP and state: I request that the appropriate deva for this project be connected to this SLG coning.

- Pan — State: I'd like to be connected with Pan.
- The White Brotherhood — State:

  I'd like to be connected with the appropriate members of the White Brotherhood working with ________ (the title/name of this project).

  If your project doesn't have a title or name, read your DDP aloud and ask for the appropriate members of the White Brotherhood to join the coning. You'll automatically get the best WB team for your project.
- Your higher self — State:

  I'd like my higher self to be connected to the coning.

  When more than one human member is attending the coning meeting, include each person's higher self in the coning as well. After including your own higher self, repeat the process for each person present. Don't try to connect them to the coning as a group. For each person, state: I'd like to connect ________'s higher self to the coning.

  Then wait 10 seconds before turning your attention to the next person.

Wait 15 seconds after activating the full coning so that your body can stabilize itself within the coning and with the team.

C. Take one dose (10–12 drops) of ETS Plus for Humans to make sure that the process of activating a coning did not affect your balance.* (Each person who has his or her higher self connected with the coning should also take one dose of ETS Plus.)

* Whenever you take a dose of ETS Plus for Humans, you are to take that dose orally. Do not try to shift this to yourself using the NS Application Process.

4. ACTIVATE YOUR DDP.

A. State/read aloud your DDP to the other members in your coning.

B. Then, state:

I'd like this DDP to be activated now.

Wait 10 seconds for the DDP to be fully activated.

5. Test ETS Plus for Soil-less Gardens to balance your soil-less garden's DDP.

A. State:

I'd like to test if an ETS Plus for SLG is needed for this DDP.

B. Ask:

How many drops of ETS Plus for SLG are needed?

Do a sequential kinesiology test.* (See the Essences Process, p. 55.) Place the correct number of drops of ETS Plus for SLG in a clean spoon.

* If no drops are needed, move on to step 6.

C. You will release the drops to your project using an NS application. To do this, state:

I would like the essence of these drops to be shifted for balancing this soil-less garden's DDP.

Hold the spoon out for 10 seconds while the shift occurs.

D. After 10 seconds, the shift will be complete. Wash the spoon. Do not try to save the remaining liquid because that ETS Plus cannot be used again.

6. Test if there is a need for ETS Plus for Soil to further balance and support your soil-less garden's DDP. (It is essential when activating a DDP to make sure the soil that supports that DDP is stabilized and balanced properly.**)

** When setting up and working with ETS for Soil in the SLG Troubleshooting Process, your coning partners will immediately focus on the soil that provides the supporting foundation for that SLG.

A. State:

I'd like to test if ETS Plus for Soil is needed for the soil that supports this soil-less garden's DDP.

B. Ask:

How many drops of ETS Plus for Soil are needed?

Do a sequential kinesiology test.*** (See the Essences Process, p. 55.) Place the correct number of drops of ETS Plus for Soil in a clean spoon.

*** If no drops are needed, move on to step 7.

C. You will release the drops to your project using an NS application. To do this, state:

I would like the essence of these drops to be shifted to the soil that supports this soil-less garden's DDP.

Hold the spoon out for 10 seconds while the shift occurs.

D. After 10 seconds, the shift will be complete. Wash the spoon. Do not save the remaining liquid because that ETS Plus cannot be used again.

7. CALIBRATE (ALIGN YOURSELF) TO YOUR PROJECT'S DDP.

NOTE: The following steps are for an individual who is starting a soil-less garden. Instructions for leading the process for others are in Appendix B: Perelandra SLG Calibration Process Steps for Groups or Teams.

A. State:

I would like to open a Calibration Process to align myself to this project's DDP.

B. State:

I request that Pan "shift forward" in the coning.

This is needed since Pan is the coning partner that actually does the balancing work during the Calibration Process.

C. Ask Pan:

How much time is needed for this calibration?

Do a sequential kinesiology test to determine how many minutes are needed. This can be 5 minutes to 30 minutes.

D. Once you get this information, tell Pan to proceed. State:

Begin the calibration now.

Relax while the calibration is going on. Allow your mind to remain comfortably focused on the intent of the calibration.

E. At the end of the time period: Give yourself a minute or so to shift your focus back to the room. Then, take one dose of ETS Plus for Humans.

F. Write down any insights, pictures or feelings that were experienced during the calibration.

8. DO A GENERAL BALANCING for the newly activated soil-less garden. For this, you will work with the SLG Troubleshooting Chart and follow those instructions. (See Chapter 4.) For Step 2 on the Troubleshooting Chart (Identify and describe issue/problem), write: "Starting Process: general balancing for this soil-less garden."

You do not need to open the coning listed in Step 1 on the Chart because you have already opened that coning for this Starting Process. Also, skip Step 9 (close coning) on the Chart. You will be closing the coning at the conclusion of the Starting Process.

9. After the general balancing is complete, close the coning. To do this, state (aloud or to yourself):

I'd like to close this SLG coning and session. I'd like the following to be disconnected from the coning:*

- My higher self

  (If other people were included in the coning, repeat this for each person. State: I'd like ________'s higher self to be disconnected.)
- The appropriate members of the White Brotherhood for this SLG
- Pan
- The Deva of ________ (project deva)

  (If you don't have a proper name/title for the devic connection, state: The project deva connected with this SLG coning.)

Wait 15 seconds.

Take one dose of ETS Plus for Humans

Each person who participated in the coning takes one dose of ETS Plus for Humans.

* No waiting time is needed between disconnecting each of the coning members when closing down a coning.

Your soil-less garden is now fully activated, balanced and stabilized, and you are calibrated to your project, its goals and intents. Wait *at least 24 hours* before opening the coning again and having your next meeting for this new soil-less garden.

If you get stuck or have any problems writing your DDP, identifying your coning team or moving through the Starting Process, call our Question Hot Line. (See our web site — www.perelandra-ltd.com — or our current catalog for the Question Hot Line phone number, days and times.)

Chapter 4

Perelandra Soil-less Garden Troubleshooting Chart

IN CHAPTER 4, I MOVE through the SLG Troubleshooting Chart, giving you pertinent information for each line. You might find it helpful to have a copy of a blank chart to refer to as you read the information. These *Companion* instructions include the new streamlining suggestions that will help give you the best, most efficient results when working with your project or goal.

Work with the SLG Troubleshooting Process to:

- provide general balance and stability for your project, and/or
- address problems that crop up in the project.

To prepare for doing an SLG Troubleshooting Process:

- Go to the bathroom.
- Drink water and eat a snack.
- Turn off the TV or radio or set up in another area where you won't have distractions.
- Get comfortable. Are you going to be warm enough, cool enough, is your chair comfortable and are the needed supplies set up in a way that will facilitate easy testing? Don't forget a bowl of nuts.

THE SLG CHART: LINE BY LINE

Project: Write the name of your project.

Date: The date you begin the chart.

Step 1. OPEN THE SLG FOUR-POINT CONING FOR YOUR PROJECT.

If more than one human is present, include each person's higher self.

Each person takes one dose of ETS Plus for Humans (10–12 drops). Wait 10 seconds.

Step 2. IDENTIFY & DESCRIBE ISSUE/PROBLEM.

Step 2 sets the focus for you and the coning for this session. It answers the question, "Why have we called this meeting?" It is important that you fully describe the issue you wish to focus on without limiting the focus. (Example: A meeting to discuss a new job opening in the company.)

Step 3. STATE INTENT.

I request that the coning focus be directed to the above stated issue/problem.

Read aloud or to yourself this full statement and don't skip over anything.

Step 4. ASK:

Is a Troubleshooting Process needed for this issue/problem?

This is a critical step and one designed to save you from wasting your time doing unnecessary testing. If you test "yes," continue on to Step 5. If you test "no," go to Step 9 and close the coning (go to p. 69 for Step 9).

Step 5. TROUBLESHOOTING LISTS.

This is where we get down to business and your coning partners tell you exactly what your project needs for the specific issue or problem you have described.

For this information, you will need to kinesiology test three lists (Environmental, Energy, Options) to find out what you are to do and in what order you are to do it. Place a small check in front of each listing that tests positive. This line is also where you will write the number indicating the order you are to do the work for the Energy and Options Lists.

Environmental Troubleshooting List

The following examples are not meant to be written in stone. They are here to give you ideas of how to think about different areas of your project that would fall under the heading "Environmental." When thinking about your project, you may need to add to or amend this list.

- New/Necessary equipment/Supplies

  Do you need to add equipment or tools, such as computers, phones, books on a certain subject, art supplies, lightning rods, surge protectors, a little troll with pink hair on your desk, an office plant, a Magic 8 Ball . . . ?

- Organize/Reorganize/Relocate

  This could include notes, filing, office space, furniture, equipment, moving to a new office, adjusting a schedule (including taking a vacation), financial planning . . .

- Clean

  This could include cleaning your office or cubicle, furniture, tools or equipment, vehicle, your ears, hiring a cleaning service . . .

- Repair/Upgrade

  This includes a computer or computer software; paint brushes or paint; any other tools, equipment or project space; communications equipment; lighting . . .

- Staff/Consultant change

  This includes hiring, firing, job requirements, salaries, renegotiating contracts . . .

- Meeting/Brainstorming/Funding/Resources

  This could include infusing support into a project, asking for input from others, signing up for a class or seminar . . .

- Additional needs

  This could be anything specific to what your SLG project might need.

If "Additional needs" tests positive and you don't know what it's referring to, ask your coning team for help and/or use the Perelandra Nature Cards for a clue about how to proceed.

If anything tests positive on the Environmental List, think about that information and make any needed changes in those areas. This List does not require a particular order for anything checked. You'll just need to address the area(s) indicated in a timely manner. If you are not sure what the information is "saying" to you, ask your coning partners for help. You'll get insight on this either during the Troubleshooting work or within a few days after closing the coning. This is also another area where Nature Cards can be helpful for pointing you in the right direction.

Energy Process Troubleshooting List*

* Before actually doing any of the Energy Processes, you will need to test the order in which to do them. (See Step 6, p. 67.)

NS Applications

For the Energy Process work, you may need to release drops of ETS Plus or essences to your project or to specific elements of your project. To do this, you will release any needed drops using an NS** application and setting up with Pan for a shift.*** Since Pan is already in your coning, the setup for an NS application is simple.

** NS = nature spirit

*** Whenever you take a dose of ETS Plus for Humans, you are to take that dose orally. Do not try to shift this to yourself using the NS Application Process.

1. State:

> I would like to set up for an NS application.
>
> Wait 5 seconds. Then ask:
>
> How many drops of ________ are needed?

Do a sequential kinesiology test to determine the number of drops needed. (See the Essences Process, p. 55.) Place any needed drops in a clean spoon. Then state:

> I ask Pan to shift the essence of these drops to ________ (the focus).

2. Hold out the spoon for 10 seconds. After 10 seconds, the shift will be complete.**** Continue with the next step of the process you are working with.

**** Clean the spoon. Record what was shifted and the number of drops on the SLG Chart.

ETS Plus for Soil-less Gardens Process

FOCUS: ETS Plus for Soil-less Gardens (SLG) was specially created to repair, balance and strengthen the order, organization and life vitality of the many areas and elements that make up a soil-less garden. In other words, this ETS Plus is like a balancing and strengthening tonic for your project. *To determine how you are to proceed, you will need to test the three Chart boxes (Test/treat as a unit, Test/treat separately, Telegraph Test). See the steps below for when to test and check these box(es).*

SUPPLIES NEEDED: The *Companion* ETS Plus for Soil-less Gardens Process steps, ETS Plus for SLG, the working SLG Troubleshooting Chart, pen/pencil, paper, spoon, paper towels, a watch or clock with a second hand.

For general balancing: If your test result is positive for ETS Plus for SLG, first set up to do a general balancing of the project.

1. Ask:

For a general balancing, does the project's *order* need ETS Plus? (Test.)

Then ask:

Does the project's *organization* need ETS Plus? (Test.)

Finally, ask:

Does the project's *life vitality* need ETS Plus? (Test.)

2. Ask:

Do I test this list as a unit? (Test.)

If you get a positive test result, check the "Test/treat as unit" box. If you get a negative result, this means you are to test and treat the order, organization and life vitality of your project separately. Check the "test/treat separately" box on the Chart.

☑ **Test and treat as a unit:** If you are to test and treat whatever is checked as a single unit, you will need to do only one test for finding out how many drops of ETS Plus for SLG are required for the general balancing. Only one application of those drops will be needed.

1. Ask:

How many drops of ETS Plus for SLG are needed?

Do a sequential kinesiology test. (See the Essences Process, p. 55.) Place the correct number of drops of ETS Plus for SLG in a clean spoon.

2. You will release them to your project using an NS application. To do this, state:

I would like the essence of these drops to be shifted for general balancing.

Hold the spoon out for 10 seconds while the shift occurs.

3. After 10 seconds, the shift will be complete. Record the number of drops you released on the SLG Chart. Wash the spoon. Do not try to save the remaining liquid because that ETS Plus cannot be used again.

☑ **Test and treat separately:** If your test result was negative to the question ("Do I test this list as a unit?"), you are to test and treat the elements — order, organization and life vitality — separately. For this, find out how many drops of ETS Plus for SLG are needed and do an NS application for *each element.*

For example, if order needs ETS Plus for SLG, find out how many drops are to be released. Then release them to the project's order using an NS application. State: I would like the essence of these drops to be shifted for general balancing to the project's order. Repeat the process for whatever else is needed for general balancing. Release the drops for the first checked element before going on to the next checked element. Repeat, if necessary, for the third checked element. Record the number of drops you released on the SLG Chart.

☑ **Telegraph testing ETS Plus for SLG:** For full balance and strength, you may need to telegraph test ETS Plus for SLG to address the order, organization and life vitality of *a specific area(s)* of your project that is related to the issue/problem written in Step 2. Telegraph testing allows you to address your project with pinpoint accuracy. When working with the SLG Troubleshooting Chart, you may easily address a telegraph test once you have completed a general balancing.

1. If you test that Telegraph Testing is needed, state:

   I'd like to set up to telegraph ETS Plus for SLG to the specific area(s) in need.*

Wait 5 seconds.

2. Then, test how many drops are needed for this application. Do a sequential kinesiology test to determine the number of drops needed.

* You do not need to identify these areas for a telegraph test. Your coning partners know what and where they are and will be working with you to set up this testing.

(See the Essences Process, p. 55.) Although you are now addressing the areas as a unit when testing for drops, in a telegraph test the proper dosage needed in each area will automatically shift to that/those area(s) when released.

3. Release the drops using an NS application. To do this, state:

> I would like the essence of these drops to be shifted to each of the areas and elements that need balancing.

Hold the spoon out for 10 seconds while the shift occurs.

4. After 10 seconds, the shift will be complete. Record the number of drops you released on the SLG Chart. Wash the spoon. Do not try to save the remaining liquid because that ETS Plus cannot be used again.

THE ETS PLUS FOR SLG PROCESS IS COMPLETE. Ask:

> Is there a waiting period before moving on to the next Step 5 process? (Test.)

If yes, ask:

> How many days? Do a sequential kinesiology test. (See the Essences Process, p. 55.)

If there is a waiting period, go to Step 9 on the Chart and close the SLG coning. Schedule to resume the Troubleshooting *within 24 hours* once the waiting period is over.*

If you don't have a waiting period, continue on to the next process you are to do in this troubleshooting.

If you don't need to do another process for Step 5, refer to Step 7 on the SLG Chart for the conclusion of the Troubleshooting Process.

RESUMING THE TROUBLESHOOTING PROCESS AFTER A WAITING PERIOD: Get your original SLG Troubleshooting Chart. Open your soil-less garden four-point coning. Read aloud the issue/problem described in Step 2. State the intent in Step 3. Now proceed with the next process in Step 5 that you are to work with.

* If you miss the 24-hour deadline for resuming the Troubleshooting Process after a waiting period, you will need to start over. You've lost an important rhythm that was being set up by the Project Coning.

There's good news and bad news with this. The good news is that the testing you have already done for that issue has had a beneficial impact and this will be reflected in your new round of testing. The bad news is that you have to start over. Had you resumed the testing after the waiting period, your remaining testing results would have been more simple because the final processes were woven in with a larger pattern. Starting over means starting over, and your new round of testing will reflect this by being more complex.

To start over, use a clean chart and copy the information you have in Step 2 from your original chart.

More often than not, the processes will all be done within one Troubleshooting session. Sometimes, one or a couple of the processes will precede the others and a waiting period will be needed before continuing the Troubleshooting with the remaining processes. This ensures that the best timing and rhythm for these processes are maintained.

ETS Plus for Soil Process

FOCUS: When using ETS Plus for Soil, the focus is on the soil that supports and stabilizes the soil-less garden and its DDP. Often, it is not apparent just what soil area is providing that support. When setting up and working with ETS for Soil in the SLG Troubleshooting Process, your coning partners will immediately focus on the soil that provides the supporting foundation for that SLG.

SUPPLIES NEEDED: The *Companion* ETS Plus for Soil Process steps, ETS Plus for Soil, the working SLG Troubleshooting Chart, pen/pencil, paper, spoon, paper towels, a watch or clock with a second hand.

1. State:

   I'd like to focus the coning on testing ETS Plus for Soil.

   Wait 5 seconds.

2. Ask:

   How many drops of ETS Plus for Soil are needed?

Do a sequential kinesiology test. (See the Essences Process, p. 55.) Place the correct number of drops in a clean spoon.

3. Release the drops using an NS application. State:

   I would like the essence of these drops to be shifted to the soil supporting ________ (name of soil-less garden) as it relates to the issue/problem stated in Step 2.

   Hold the spoon out for 10 seconds while the shift occurs.

4. After 10 seconds, the shift will be complete. Record the number of drops you released on the SLG Chart. Wash the spoon. Do not try to save the remaining liquid because that ETS Plus cannot be used again.

5. That's it. This process is complete. Ask:

   Is there a waiting period before moving on to the next Step 5 process? (Test.)

   If yes, ask:

      How many days? Do a sequential kinesiology test. (See the Essences Process, p. 55.)

If there is a waiting period, go to Step 9 on the Chart and close the SLG coning. Schedule to resume the Troubleshooting *within 24 hours* once the waiting period is over. If you don't resume the Troubleshooting within this time period, see p. 39.

If you don't have a waiting period, continue on to the next process you are to do in this troubleshooting.

If you don't need to do another process for Step 5, refer to Step 7 on the SLG Chart for the conclusion of the Troubleshooting Process.

RESUMING THE TROUBLESHOOTING PROCESS AFTER A WAITING PERIOD: Get your original SLG Troubleshooting Chart. Open your soil-less garden four-point coning. Read aloud the issue/problem described in Step 2. State the intent in Step 3. Now proceed with the next process in Step 5 that you are to work with.

More often than not, the processes will all be done within one Troubleshooting session. Sometimes, one or a couple of the processes will precede the others and a waiting period will be needed before continuing the Troubleshooting with the remaining processes. This ensures that the best timing and rhythm for these processes are maintained.

ADDITIONAL USES FOR ETS PLUS FOR SOIL WITH SLGS

The following is a list of other ways ETS Plus for Soil may be used with soil-less gardens. For more information and the steps, see the "ETS Plus for Soil" brochure that is included with each order for this ETS Plus. The brochure may also be read/downloaded from our web site. Use ETS Plus for Soil

- During times of soil-less garden distress
- For general strengthening
  - For an already-activated soil-less garden and DDP
  - Short-term soil-less gardens

Energy Cleansing Process

THE FOCUS of this Energy Cleansing Process is your project as it relates to the issue/problem that you described in Step 2.

SUPPLIES NEEDED: The *Companion* Energy Cleansing Process steps, ETS Plus for SLG, ETS Plus for Soil, the working SLG Troubleshooting Chart, pen/pencil, paper, spoon, paper towels, a watch or clock with a second hand.

1. To determine the area to be cleansed, test the three boxes: Project Framework (the physical aspects of your project such as office, building, camera, etc.), Mental-Level Activity (the thinking, planning, considering, daydreaming, idea-producing activities that your project requires), Both (both).

Whatever tests positive is the area(s) you will be working with. For most projects it will be difficult to accurately draw a sketch of these three areas. So let's make this easy. Get a sheet of paper and a pen/pencil/crayon/paintbrush of your choice.

- If "Project Framework" tested positive, draw a circle.* Inside that circle write "Project Framework."

- If "Mental-Level Activity" tested positive, draw a circle and inside write "Mental-Level Activity."

- If "Both" tested positive, draw a larger circle and inside write "Project Framework and Mental-Level Activity."

* HELP FOR THE DRAWING IMPAIRED: Are you overwhelmed at the idea of drawing a circle? I am providing a circle template for you to photocopy and use. You'll find it in the back of the book with the blank SLG Charts.

Although your focus will be on this labeled circle throughout the Process, the actual energy cleansing will occur within your project according to the label, no matter how large or extensive the project. This same principle would apply even if you were working with a global project with different office locations. For you to do this Process comfortably, the circle says it all.

> (Warning: Do not call Perelandra to ask how big the circles should be! That will just annoy us. Make them big enough so that you can easily write these words!)

2. Prepare yourself for doing the Energy Cleansing Process. Drink some water, relax and get comfortable. Make sure your labeled circle can be easily seen.

3. Directing your attention to the coning, state to yourself or aloud:

> I would like to do the Energy Cleansing Process for the issue/problem as it relates to this soil-less garden.

4. Then state:

> In light of that focus and intent, I ask that any inappropriate, stagnant, darkened or ungrounded energies be released from this soil-less garden. I request this knowing that the process I am about to be a part of is an evolutionary process of life.

5. Visualize or look at the area to be cleansed (the labeled circle).

6. Visualize a white sheet of light forming 3 feet below where the paper with the circle is sitting. Allow the outside edges of the sheet of light to extend 3 feet beyond the edges of the paper.

7. State:

> I ask the members of this SLG coning to join me as together we slowly move the sheet up and through the area to be cleansed.

"Watch" or sense the sheet as it moves. For you, it will be moving up to and through the circle and its surrounding area. Allow the sheet to rise 3 feet above the circle/paper, and then stop.

If the sheet moves too slowly or too quickly, or one side is not rising to the same height, ask that the sheet stop; request any changes and then ask that the sheet continue to be raised. The changes you requested will be in place. Remember: You do not need to give yourself a hernia raising this sheet. You have the entire coning team helping you. If you relax, the sheet will move smoothly, easily and effortlessly.

8. State:

> I would now like to create a bundle with the sheet.

Visualize or sense the sheet forming a bundle of white light that totally encloses the collected energies. To the left of the bundle, see a gold cord. Tie the bundle closed with this cord. If you do not "see" a gold cord, simply request that the coning help you by tying the bundle closed with a gold cord.

9. State:

> I call for the release of the bundle, so that the energies that have been gathered can continue with their own evolutionary process.

Watch the bundle lift. IMPORTANT: Just watch. Don't try to determine where the bundle is going or what is happening to it. Your coning partners take care of this.

10. Return your focus to the now-cleansed circle for one minute. Observe any changes you might feel.

Most people sense or feel nothing. If this is the case for you, don't fret. Nothing went wrong. And there's nothing wrong with you either. Just sit quietly for one minute before continuing with the rest of the process.

11. Shift your attention to your breathing, focusing on your body as you inhale and exhale 3 or 4 times.

12. Thank your SLG team for helping you with a job well done. Also acknowledge the cooperation of the white sheet, the gold cord and the energies that were released.

13. Return your attention to your project circle.

14. Balance and stabilize the work that was just completed.

BALANCING:

For balancing, work with ETS Plus for Soil.

A. State:

I'd like to test if ETS Plus for Soil is needed for any balancing due to this energy cleansing work.

B. Ask:

How many drops of ETS Plus for Soil are needed?

Do a sequential kinesiology test.* (See the Essences Process, p. 55.) Place the correct number of drops of ETS Plus for Soil in a clean spoon.

* If no drops are needed, move on to stabilizing.

C. You will release them to your project using an NS application.

To do this, state:

I would like the essence of these drops to be shifted for balancing my project in light of the Energy Cleansing Process work.

Hold the spoon out for 10 seconds while the shift occurs.

D. After 10 seconds, the shift will be complete. Record the number of drops you released on the SLG Chart. Wash the spoon. Do not try to save the remaining liquid because that ETS Plus cannot be used again.

STABILIZING:

For stabilizing, work with ETS Plus for Soil-less Gardens.

A. State:

I'd like to test if ETS Plus for SLG is needed for any stabilizing due to the energy cleansing work.

B. Ask:

How many drops of ETS Plus for SLG are needed?

Do a sequential kinesiology test.** (See the Essences Process, p. 55.) Place the correct number of drops of ETS Plus for SLG in a clean spoon.

** If no drops are needed, move on to the next step.

C. You will release them to your project using an NS application. To do this, state:

> I would like the essence of these drops to be shifted for stabilizing my project in light of the Energy Cleansing Process work.

Hold the spoon out for 10 seconds while the shift occurs.

D. After 10 seconds, the shift will be complete. Record the number of drops you released on the SLG Chart. Wash the spoon. Do not try to save the remaining liquid because that ETS Plus cannot be used again.

15. This process is complete. Ask:

Is there a waiting period before moving on to the next Step 5 process? (Test.)

If yes, ask:

> How many days? Do a sequential kinesiology test. (See the Essences Process, p. 55.)

If there is a waiting period, go to Step 9 on the Chart and close the SLG coning. Schedule to resume the Troubleshooting *within 24 hours* once the waiting period is over. If you don't resume the Troubleshooting within this time period, see p. 39.

If you don't have a waiting period, continue on to the next process you are to do in this troubleshooting.

If you don't need to do another process for Step 5, refer to Step 7 on the SLG Chart for the conclusion of the Troubleshooting Process.

RESUMING THE TROUBLESHOOTING PROCESS AFTER A WAITING PERIOD: Get your original SLG Troubleshooting Chart. Open your soil-less garden four-point coning. Read aloud the issue/problem described in Step 2. State the intent in Step 3. Now proceed with the next process in Step 5 that you are to work with.

More often than not, the processes will all be done within one Troubleshooting session. Sometimes, one or a couple of the processes will precede the others and a waiting period will be needed before continuing the Troubleshooting with the remaining processes. This ensures that the best timing and rhythm for these processes are maintained.

Battle Energy Release Process

THE FOCUS of the Battle Energy Release Process is your project as it relates to the issue/problem that you described in Step 2.

SUPPLIES NEEDED: The *Companion* Battle Energy Release Process steps, ETS Plus for SLG, ETS Plus for Soil, the working SLG Troubleshooting Chart, pen/pencil, paper, spoon, paper towels, a watch or clock with a second hand.

1. State:

I would like to do the Battle Energy Release Process for ________ (your project) as it relates to the issue/problem described in Step 2 and the intent and focus of this chart. I want to do this work in the spirit of care, concern and co-creativity to help release the appropriate battle energies.

2. The release. State:

I call for the release of all the battle energies in question with gentleness and ease. I request that the released energies gather as an "energy cloud" 3 feet above my head.

Wait 15 minutes for the release to complete.

If at any time you are uncomfortable while an energy release is going on, simply ask that the release occur a little more slowly or gently — or whatever is needed for you to feel comfortable. The members of the coning will immediately comply with your request and adjust the release.*

3. When the release is complete, state:

I ask that the battle energy that has gathered as an energy cloud over my head now release and continue on with its evolutionary process.

Do not try to determine where or what that might be.

* During the release you may feel or sense sensations in and around your body. All you need to do is relax and let the process continue its work. Most people feel nothing. If this is the case for you, nothing is wrong — with you or the process. Just wait quietly for the full 15 minutes while the release is going on.

* Many people feel nothing. If this is the case for you, nothing is wrong. Just focus on your project and wait quietly for 2 minutes as the grounding takes place.

4. Spend 2 minutes noticing the changes and sensations you might be feeling. This serves to fully ground the completed process for you.*

5. Balance and stabilize the work you have just completed. The balancer and the stabilizer are for those elements of your project that have been impacted by the Battle Energy Release Process.

BALANCING:

For balancing, work with ETS Plus for Soil.

A. State:

I'd like to test if ETS Plus for Soil is needed for any balancing due to this Battle Energy Release work.

B. Ask:

How many drops of ETS Plus for Soil are needed?

** If no drops are needed, move on to stabilizing.

Do a sequential kinesiology test.** (See the Essences Process, p. 55.) Place the correct number of drops of ETS Plus for Soil in a clean spoon.

C. You will release them to your project using an NS application. To do this, state:

I would like the essence of these drops to be shifted for balancing my project in light of the Battle Energy Release Process work.

Hold the spoon out for 10 seconds while the shift occurs.

D. After 10 seconds, the shift will be complete. Record the number of drops you released on the SLG Chart. Wash the spoon. Do not try to save the remaining liquid because that ETS Plus cannot be used again.

STABILIZING:

For stabilizing, work with ETS Plus for Soil-less Gardens.

A. State:

I'd like to test if ETS Plus for SLG is needed for any stabilizing due to this Battle Energy Release work.

B. Ask:

How many drops of ETS Plus for SLG are needed?

Do a sequential kinesiology test.* (See the Essences Process, p. 55.) Place the correct number of drops of ETS Plus for SLG in a clean spoon.

C. You will release them to your project using an NS application. To do this, state:

> I would like the essence of these drops to be shifted for stabilizing my project in light of the Battle Energy Release Process work.

Hold the spoon out for 10 seconds while the shift occurs.

D. After 10 seconds, the shift will be complete. Record the number of drops you released on the SLG Chart. Wash the spoon. Do not try to save the remaining liquid because that ETS Plus cannot be used again.

THE CHART

Battle Energy Release Process

* If no drops are needed, move on to the next step.

6. This process is complete. Ask:

Is there a waiting period before moving on to the next Step 5 process? (Test.)

If yes, ask:

> How many days? Do a sequential kinesiology test. (See the Essences Process, p. 55.)

If there is a waiting period, go to Step 9 on the Chart and close the SLG coning. Schedule to resume the Troubleshooting *within 24 hours* once the waiting period is over. If you don't resume the Troubleshooting within this time period, see p. 39.

If you don't have a waiting period, continue on to the next process you are to do in this troubleshooting.

If you don't need to do another process for Step 5, refer to Step 7 on the SLG Chart for the conclusion of the Troubleshooting Process.

RESUMING THE TROUBLESHOOTING PROCESS AFTER A WAITING PERIOD: Get your original SLG Troubleshooting Chart. Open your soil-less garden four-point coning. Read aloud the issue/problem described in Step 2. State the intent in Step 3. Now proceed with the next process in Step 5 that you are to work with.

More often than not, the processes will all be done within one Troubleshooting session. Sometimes, one or a couple of the processes will precede the others and a waiting period will be needed before continuing the Troubleshooting with the remaining processes. This ensures that the best timing and rhythm for these processes are maintained.

Balancing and Stabilizing Process

TEST THE FOLLOWING to determine the focus(es):

- general balancing/stabilizing for the full project, and/or
- balancing/stabilizing specific elements related to the issue/problem described in Step 2. Sometimes both will test positive. In that case, you are to do the process for each focus. Test them independently for the order in which they are to be done.* (They might not be done together or in consecutive order.)

* Ask: Do I do the general balancing for the full project first? (Test.) If "yes," you now know you are to work with the full project first. If "no," you are to balance and stabilize specific elements first.

SUPPLIES NEEDED: The *Companion* Balancing and Stabilizing Process steps, ETS Plus for SLG, ETS Plus for Soil, the working SLG Troubleshooting Chart, pen/pencil, paper, spoon, paper towels, a watch or clock with a second hand.

1. State:

I would like to do the SLG Balancing and Stabilizing Process for the focus: ________.

(Read the focus that you need to balance and stabilize.)

2. BALANCING:

For balancing, work with ETS Plus for Soil.

A. State:

I'd like to test if ETS Plus for Soil is needed for balancing.

B. Ask:

How many drops of ETS Plus for Soil are needed?

Do a sequential kinesiology test.** (See the Essences Process, p. 55.) Place the correct number of drops of ETS Plus for Soil in a clean spoon.

** If no drops are needed, move on to stabilizing.

C. You will release them to your project using an NS application.

To do this, state:

> I would like the essence of these drops to be shifted for balancing my project.

Hold the spoon out for 10 seconds while the shift occurs.

D. After 10 seconds, the shift will be complete. Record the number of drops you released on the SLG Chart. Wash the spoon. Do not try to save the remaining liquid because that ETS Plus cannot be used again.

3. STABILIZING:

For stabilizing, work with ETS Plus for Soil-less Gardens.

A. State:

> I'd like to test if ETS Plus for SLG is needed for stabilizing.

B. Ask:

> How many drops of ETS Plus for SLG are needed?

Do a sequential kinesiology test.* (See the Essences Process, p. 55.) Place the correct number of drops of ETS Plus for SLG in a clean spoon.

C. You will release them to your project using an NS application.

To do this, state:

> I would like the essence of these drops to be shifted for stabilizing my project.

Hold the spoon out for 10 seconds while the shift occurs.

D. After 10 seconds, the shift will be complete. Record the number of drops you released on the SLG Chart. Wash the spoon. Do not try to save the remaining liquid because that ETS Plus cannot be used again.

4. Spend 2 minutes quietly sensing any changes.** You do not need to further balance and stabilize the Balancing and Stabilizing Process!

5. This process is complete. Ask:

> Is there a waiting period before moving on to the next Step 5 process? (Test.)

* If no drops are needed, move on to the next step.

** Many people feel nothing. If this is the case for you, nothing is wrong. Just focus on your project and wait quietly for 2 minutes as the grounding takes place.

If yes, ask:

How many days? Do a sequential kinesiology test.
(See the Essences Process, p. 55.)

If there is a waiting period, go to Step 9 on the Chart and close the SLG coning. Schedule to resume the Troubleshooting *within 24 hours* once the waiting period is over. If you don't resume the Troubleshooting within this time period, see p. 39.

If you don't have a waiting period, continue on to the next process you are to do in this troubleshooting.

If you don't need to do another process for Step 5, refer to Step 7 on the SLG Chart for the conclusion of the Troubleshooting Process.

RESUMING THE TROUBLESHOOTING PROCESS AFTER A WAITING PERIOD: Get your original SLG Troubleshooting Chart. Open your soil-less garden four-point coning. Read aloud the issue/problem described in Step 2. State the intent in Step 3. Now proceed with the next process in Step 5 that you are to work with.

More often than not, the processes will all be done within one Troubleshooting session. Sometimes, one or a couple of the processes will precede the others and a waiting period will be needed before continuing the Troubleshooting with the remaining processes. This ensures that the best timing and rhythm for these processes are maintained.

Essences Process

FOCUS: Remember that you are not taking these essence solutions for yourself — you are releasing them to your project. For this, you will be releasing the essences using NS applications only. You will not be taking them orally.

> IMPORTANT: Whenever administering an essence solution for a soil-less garden, remember to focus on what the solution is for at the time you are releasing it. In the Step 5 Essences Process, if you are administering a general solution, the focus is on your project as a whole as it relates to the issue being addressed. When administering a solution for

a specific problem, the focus is the elements of the project that are involved with the specific problem being addressed. You need not identify the problem any further because it has already been identified by the coning.

SUPPLIES NEEDED: The *Companion* Essences Process steps, the Perelandra Rose and Rose II Essences,* the working SLG Troubleshooting Chart, pen/pencil, paper, spoon, paper towels, a watch or clock with a second hand.

* When testing the Essences Process, you may include the other sets of Perelandra Essences, if you wish.

1. Test to determine whether you are doing a general solution and/or a solution for a specific problem. Check the appropriate line on the SLG Chart. Then state your intent to essence test for whatever tested positive. If both tested positive, you will do the testing in the following order:

FIRST: Test the specific problem and administer those essences.

Focus on the words "specific problem" as you are releasing the essences.

SECOND: Test the general solution and administer those essences.

Focus on the words "general solution" as you are releasing these essences.

2. Place each box of essences, one at a time, in your lap. Ask:

Are any essences from this box needed for ________? (the specific focus for this test) (Test.)

If you get a negative, you don't have to test the bottles from that box because none are needed.

If you get a positive, test each bottle from that box individually to determine which ones are needed. For each bottle, ask:

Is ________ Essence needed? (Test.)

3. Check your results by placing in your lap all the bottles that tested positive. Ask:

Are these the only essences needed? (Test.)

If the response is positive, go to step 4. (If only one essence is needed, skip step 4 and go to step 5.)

If you get a negative, retest the other essences. A negative means you missed an essence and need to find what was missed. After retesting, ask the question again:

Are these the only essences needed? (Test.)

If the response is still negative, keep testing the essences until you get a positive response to the question. This will verify that you have all the essences required for the process.

4. If you have more than one essence, check them as a combination by placing all of them in your lap and asking:

Is this the combination needed for this soil-less garden? (Test.)

If you get a negative even though the essences tested positive when tested individually, you may need to adjust the combination.

Place each of the combination bottles in your lap separately and ask:

Do I remove this bottle from the combination? (Test.)

There is no shortcut for checking a step 4 essence combination. The bottles that test positive must be taken out of the box(es) and placed in your lap in order to test them properly as a combination.

Whatever tests positive gets removed. Then put the remaining combination bottles in your lap and ask:

Is this combination now correct? (Test.)

This means that when the individual essences that tested positive were put together, a combination was created that made one or more of those essences unnecessary. The whole was stronger and more effective than the sum of its parts.

You should get a positive. If you don't, test the *original combination* again, and keep working at it until they test positive as a unit.

5. Administer the essences using an NS application. Remember, you are administering them for the project and the areas specifically set up for this testing. Record the essences on the SLG Chart.

NS APPLICATION FOR ESSENCES

For this, you will be setting up for a Pan shift. After testing the Perelandra Essences, put *one drop* of each needed essence in a spoon.

1. State:

> I would like to set up for an NS application.
>
> Then state:
>
> I ask Pan to shift these essences to any area(s) that has been set up as the focus of this process.

2. Hold out the spoon with the essences for 10 seconds. Clean the spoon. Record the essences on the SLG Chart.

Testing for Dosage

Because of the nature of soil-less gardens, most essence solutions will be needed one time only. But it is always wise to check for dosage (how many days/weeks you are to administer the essences), just in case.

1. Hold in your lap all the bottles that tested positive as a unit.

2. Ask:

> Are these essences needed more than one time? (Test.)

If negative, that means, "No, they don't need to be administered more than one time," and you have already completed step 5. (See p. 54.)

3. If positive, find out how many days you should administer them and how many times per day.

SEQUENTIAL TEST STEPS: Do a sequential test to determine the number of days. With the needed essences in your lap, ask:

> Are they needed one day? (Test.)
>
> 2 days? (Test.)
>
> 3 days? (Test.)
>
> Do a count until you get a negative response.

If you need to administer the essences for 3 days, you will test positive when you ask, "one day?", "2 days?", "3 days?" When you ask "4 days?" you will test negative. That will tell you that your project would be assisted and strengthened by having the essences released to it for 3 days, not 4 days.

Daily Dosage

1. Using the same format, ask:

   Should these essences be administered to the focus of this process one time daily? (Test.)

   2 times daily? (Test.)

And so on, until you get a negative. Your last positive will tell you how many times per day they are needed. Record the dosage on the SLG Chart.

2. Generally, essences are to be administered in the morning and/or in the evening and/or in mid-afternoon. If you wish to be more precise, test to see if it is best to administer them in the morning, afternoon or evening, or any combination of the three. Ask:

   Should these essences be administered in the morning? (Test.)

   In the afternoon? (Test.)

   In the evening? (Test.)

Whatever tests positive is when you should administer one of the dosages.*

* If you end up with more positives for times of day than you need, retest. If the times of day (morning, noon, evening) are still not in sync with the number of times per day (2 times daily), you can forget this and just pick when you'd like to release the dosages. Just make sure the rhythm you pick for one day (morning and evening) is the same rhythm you use throughout the time you are to release this particular solution.

3. This process is complete. Ask:

   Is there a waiting period before moving on to the next Step 5 process? (Test.)

   If yes, ask:

   How many days? Do a sequential kinesiology test.

If there is a waiting period, go to Step 9 on the Chart and close the SLG coning. Schedule to resume the Troubleshooting *within 24 hours* once the waiting period is over. If you don't resume the Troubleshooting within this time period, see p. 39.

If you don't have a waiting period, continue on to the next process you are to do in this troubleshooting.

If you don't need to do another process for Step 5, refer to Step 7 on the SLG Chart for the conclusion of the Troubleshooting Process.

RESUMING THE TROUBLESHOOTING PROCESS AFTER A WAITING PERIOD: Get your original SLG Troubleshooting Chart. Open your soil-less garden four-point coning. Read aloud the issue/problem described in Step 2. State the intent in Step 3. Now proceed with the next process in Step 5 that you are to work with.

More often than not, the processes will all be done within one Troubleshooting session. Sometimes, one or a couple of the processes will precede the others and a waiting period will be needed before continuing the Troubleshooting with the remaining processes. This ensures that the best timing and rhythm for these processes are maintained.

Making a Solution Bottle for SLG Projects Only

Make an essence solution bottle for your project — 4 drops of each essence per half-ounce bottle — if it is to be released more than a couple of days or if you need to release it several times throughout the day and must carry the solution around. Add a teaspoon of brandy or distilled white vinegar if you need to preserve the solution, and fill with spring or untreated water. If this is unavailable, tap water will do. You can also refrigerate the solution, thus eliminating the need for preserving it with brandy or vinegar. To release the solution drops to your project, open the coning, focus on the intent of the Essences Process (general solution or specific issue) and on the intent of the work (Step 2 on the Chart), then release *one dropperful* (10–12 drops) for each dosage using an NS application. Then close the coning.

Follow-up Testing for the Step 5 Essences Process

IMPORTANT: Remember to do follow-up essence testing *within 36 hours* once the dosage for each essence solution is completed.

1. Open the SLG coning. Take one dose of ETS Plus for Humans.

2. Read Steps 2 and 3 on the SLG Chart you are working with. This reestablishes the focus and intent of the testing.

3. State:

   I would like to do a follow-up test for the Essences Process for this SLG focus.

Test the Perelandra Essences and check for a new dosage. If you test that no other essences are needed, your project is clear and you do not need to do any further testing for that Essences Process. Close the SLG coning (Step 9).

If a new solution was needed, be sure to do another follow-up test *within 36 hours* once the new dosage has been completed. Each time you release a dose of a new solution (using an NS application), you will need to think about the intent of the work. Continue the follow-up testing until you test clear and your project no longer needs any essences for this particular focus.

Triangulation Process

THE FOCUS of this SLG Triangulation Process is the issue/problem described in Step 2.

SUPPLIES NEEDED: The *Companion* Triangulation Process steps, ETS Plus for SLG, ETS Plus for Soil, the working SLG Troubleshooting Chart, pen/pencil, paper, spoon, paper towels, a watch or clock with a second hand.

1. State:

   I would like to do the SLG Triangulation Process as it relates to the issue/problem described in Step 2 and to the project as a whole.

At this point, your coning partners will identify the first triangle and any additional triangles for you. In order to do this process, you do not need to know where the points of any triangles are located. Figuring this out would be wasted time and brain energy on your part. Your partners also know exactly what point(s) is in need.

2. Identify which point(s) and/or link(s) of a triangle are involved by focusing on each of the boxes on the SLG Chart, one at a time, and asking:

Does point/link #__ need balancing and stabilizing? (Test.)

Whatever tests positive is a weak spot. Check the box(es) that tested positive. You may test positive for more than one weak spot in a triangle.

3. Do a balancing and stabilizing for each weak point and link and use an NS application for administering the balancer and stabilizer. For each weak point/link, do the following:

BALANCING:

For balancing, work with ETS Plus for Soil.

A. State:

I'd like to test if ETS Plus for Soil is needed to balance point/link #__.

B. Ask:

How many drops of ETS Plus for Soil are needed?

Do a sequential kinesiology test.* (See the Essences Process, p. 55.) Place the correct number of drops of ETS Plus for Soil in a clean spoon.

* If no drops are needed, move on to stabilizing.

C. Release them to the specific point/link using an NS application. To do this, state:

I would like the essence of these drops to be shifted for balancing point/link #__.

Hold the spoon out for 10 seconds while the shift occurs.

D. After 10 seconds, the shift will be complete. Record the number of drops you released on the SLG Chart. Wash the spoon. Do not try to save the remaining liquid because that ETS Plus cannot be used again.

Stabilizing:

For stabilizing, work with ETS Plus for Soil-less Gardens.

A. State:

I'd like to test if ETS Plus for SLG is needed to stabilize point/link #__.

B. Ask:

How many drops of ETS Plus for SLG are needed?

Do a sequential kinesiology test.* (See the Essences Process, p. 55.) Place the correct number of drops of ETS Plus for SLG in a clean spoon.

* If no drops are needed, move on to the next step.

C. Release them to your project using an NS application.

To do this, state:

I would like the essence of these drops to be shifted for stabilizing point/link #________.

Hold the spoon out for 10 seconds while the shift occurs.

D. After 10 seconds, the shift will be complete. Record the number of drops you released on the SLG Chart. Wash the spoon. Do not try to save the remaining liquid because that ETS Plus cannot be used again.

Important: Do both the balancing and stabilizing for each weak point or link before addressing the next weak point or link.

4. Once the work for the first triangle is complete, ask:

Is there another triangle I need to work with? (Test.)

If so, repeat steps 2 and 3 for each triangle in need of balancing and stabilizing. Set up the testing and record any additional triangle information on the back of your working SLG Chart.

5. Once all triangles have been addressed, this process is complete. Ask:

Is there a waiting period before moving on to the next Step 5 process? (Test.)

If yes, ask:

How many days? Do a sequential kinesiology test. (See the Essences Process, p. 55.)

If there is a waiting period, go to Step 9 on the Chart and close the SLG coning. Schedule to resume the Troubleshooting *within 24 hours* once the waiting period is over. If you don't resume the Troubleshooting within this time period, see p. 39.

If you don't have a waiting period, continue on to the next process you are to do in this troubleshooting.

If you don't need to do another process for Step 5, refer to Step 7 on the SLG Chart for the conclusion of the Troubleshooting Process.

RESUMING THE TROUBLESHOOTING PROCESS AFTER A WAITING PERIOD: Get your original SLG Troubleshooting Chart. Open your soil-less garden four-point coning. Read aloud the issue/problem described in Step 2. State the intent in Step 3. Now proceed with the next process in Step 5 that you are to work with.

More often than not, the processes will all be done within one Troubleshooting session. Sometimes, one or a couple of the processes will precede the others and a waiting period will be needed before continuing the Troubleshooting with the remaining processes. This ensures that the best timing and rhythm for these processes are maintained.

SLG Calibration Process

FOCUSES: Test this list to determine the focus for the SLG Calibration Process.*

- To align an individual or team with a project's DDP
- To add a new member to a project's team and align the new team with the project's DDP (see p. 125)
- To remove a member from the team, re-form the new team and align this new team with the project's DDP (see p. 125)
- To realign an individual or team with the project's DDP during times of development or trouble
- To align the full project or the specific problem/issue to the DDP

* Check the box on the SLG Calibration Chart line that tests positive. If more than one box tests positive, you will need to do an SLG Calibration Process for each one. Test to find out which focus you are to calibrate first, second, and so on.

Supplies needed: The *Companion* SLG Calibration Process steps, ETS Plus for SLG, ETS Plus for Soil, ETS Plus for Humans, the working SLG Troubleshooting Chart, pen/pencil, paper, spoon, paper towels, a watch or clock with a second hand.

> *Note: The following steps are for an individual who is working with a soil-less garden. The instructions for leading the process for others are in Appendix B: Perelandra SLG Calibration Process Steps for Groups or Teams.*

1. State:

> I would like to open an SLG Calibration Process for ________.
> (the focus for this calibration)

2. State:

> I request that Pan "shift forward" in the coning.

This is needed since Pan is the coning partner that actually does the balancing work during the Calibration Process.

3. Ask Pan:

> How much time is needed for this calibration?

Do a sequential kinesiology test to determine how many minutes are needed. (See the Essences Process, p. 55.) This can be 5 minutes to 30 minutes.

4. Once you get this information, tell Pan to proceed. State:

> Begin the calibration now.
> Relax while the calibration is going on. Allow your mind to remain comfortably focused on the intent of the calibration.

5. At the end of the time period: Shift your focus back to the room for 2 minutes. Then, take one dose (10–12 drops) of ETS Plus for Humans to restore any needed balance as a result of the calibration.

6. Balance and stabilize the calibration work with the project.

Balancing:

For balancing, work with ETS Plus for Soil.

A. State:

I'd like to test if ETS Plus for Soil is needed to balance this calibration.

B. Ask:

How many drops of ETS Plus for Soil are needed?

Do a sequential kinesiology test.* (See the Essences Process, p. 55.) Place the correct number of drops of ETS Plus for Soil in a clean spoon.

* If no drops are needed, move on to stabilizing.

C. Release them according to the focus of the calibration using an NS application.

To do this, state:

I would like the essence of these drops to be shifted for balancing this SLG Calibration Process for ________. (Restate the focus.)

Hold the spoon out for 10 seconds while the shift occurs.

D. After 10 seconds, the shift will be complete. Record the number of drops you released on the SLG Chart. Wash the spoon. Do not try to save the remaining liquid because that ETS Plus cannot be used again.

STABILIZING:

For stabilizing, work with ETS Plus for Soil-less Gardens.

A. State:

I'd like to test if ETS Plus for SLG is needed to stabilize the calibration.

B. Ask:

How many drops of ETS Plus for SLG are needed?

Do a sequential kinesiology test.** (See the Essences Process, p. 55.) Place the correct number of drops of ETS Plus for SLG in a clean spoon.

** If no drops are needed, move on to the next step.

C. Release them according to the focus of the calibration using an NS application.

To do this, state:

I would like the essence of these drops to be shifted for stabilizing the calibration for ________. (Restate the focus.)

Hold the spoon out for 10 seconds while the shift occurs.

D. After 10 seconds, the shift will be complete. Record the number of drops you released on the SLG Chart. Wash the spoon. Do not try to save the remaining liquid because that ETS Plus cannot be used again.

7. Make note of any insights, pictures or feelings that were experienced during the calibration.

8. Once each needed focus has been calibrated, this process is complete. Ask:

Is there a waiting period before moving on to the next Step 5 process? (Test.)

If yes, ask:

How many days? Do a sequential kinesiology test. (See the Essences Process, p. 55.)

If there is a waiting period, go to Step 9 on the Chart and close the SLG coning. Schedule to resume the Troubleshooting *within 24 hours* once the waiting period is over. If you don't resume the Troubleshooting within this time period, see p. 39.

If you don't have a waiting period, continue on to the next process you are to do in this troubleshooting.

If you don't need to do another process for Step 5, refer to Step 7 on the SLG Chart for the conclusion of the Troubleshooting Process.

RESUMING THE TROUBLESHOOTING PROCESS AFTER A WAITING PERIOD: Get your original SLG Troubleshooting Chart. Open your soil-less garden four-point coning. Read aloud the issue/problem described in Step 2. State the intent in Step 3. Now proceed with the next process in Step 5 that you are to work with.

More often than not, the processes will all be done within one Troubleshooting session. Sometimes, one or a couple of the processes will precede the others and a waiting period will be needed before continuing the Troubleshooting with the remaining processes. This ensures that the best timing and rhythm for these processes are maintained.

SLG Troubleshooting Process Options

Here are some options for you to consider for Step 5 of the SLG Troubleshooting Process and Chart. Each of these options could be important to reestablishing the balance needed in your project. However, in order to add them to the Chart's Step 5 list for testing, you need to decide which of the options you wish to work with. Once you've decided this, include those lines in your regular Step 5 testing to determine if any are needed for the intent of the work.

Human Troubleshooting List

This list will let you know if *you* are part of the problem that has been described in Step 2 and what you need to do to balance yourself in this situation. If the first two options in this list test positive, include them when determining the order (Step 6) you are to do the processes. They are easy and not time consuming, so slipping them in with the Troubleshooting order that's determined in Step 6 will be easy.

1. ETS PLUS FOR HUMANS: If this tests positive, read or state aloud what you have written in Step 2. Then take one dose (10–12 drops) of ETS Plus for Humans.*
2. ESSENCES TELEGRAPH TEST: The focus for this telegraph test is what is written in Step 2. When setting up for the test, read or state aloud that information. This telegraph test addresses *your* personal issues that relate to the intent of the SLG Troubleshooting. (To telegraph test, see the Quick-Glance Essence Telegraph Test Steps.**)

If any of the next three options in this list test positive, test the order you are to do them in and schedule to do them *within 24 hours after closing* the SLG Troubleshooting Process and Chart. Set up each human process by reading or stating aloud what is written in Step 2 on the SLG Chart.***

3. PERSONAL CALIBRATION PROCESS (See *MAP, Third Edition* for steps.)
4. MAP: Read the Step 2 information to your MAP team. Talk to them about any thoughts, feelings, issues, fears, insights and concerns you are having that relate to the issue/problem in Step 2. (See *MAP, Third Edition* for the steps.)

* Whenever you take a dose of ETS Plus for Humans, you are to take that dose orally. Do not try to shift this to yourself using the NS Application Process.

** *Quick-Glance Essence Telegraph Test Steps:*
1. To prepare, do a basic essence test. Take one drop of each needed essence. Do not test for dosage.
2. State or read aloud the focus of your telegraph test written in Step 2.
3. Do a second basic essence test. Take one drop of each needed essence. Test for dosage (see p. 55) for the number of days you are to take these essences.
4. Each time you take a dose of these essences, first state or read the telegraphing focus. Then take one drop of each needed essence. You do not need to open a coning.

For more detailed information about essences telegraph testing see the book, *Flower Essences.*

*** See pp. 67–68 for more ordering information and how to schedule multiple personal processes.

5. MAP/CALIBRATION: The focus of the calibration is yourself in relation to the SLG issue/problem described in Step 2. Set up in MAP just as you would any other MAP/Calibration. Talk about any thoughts, feelings, issues, fears, insights and concerns you are having that relate to the issue/problem. (See *MAP, Third Edition* for the steps.)

Workbook II Troubleshooting Process & Chart

Sometimes the balance of your project requires that something in the project's natural environment be addressed. Should this line on the SLG Chart test positive, that is an area in need of work.

- For this, you will be using the *Perelandra Garden Workbook II* and the *Workbook II* Troubleshooting Chart.

- *Within 24 hours* after completing the SLG Troubleshooting Process and Chart, turn your attention to *Workbook II* and its Chart.

- For Step 4 on the *Workbook II* Chart, write the name of your SLG project and copy what you wrote in Step 2 on the SLG Chart. This will align any *Workbook II* processes and testing with the original SLG work and intent.

- Add your project deva to this new coning. Then proceed with the *Workbook II* Chart and processes using the language as written for the *Workbook II* Troubleshooting Process.

Microbial Balancing Program Troubleshooting Process & Chart

If this line tests positive for you, it means that your project or specific elements in your project as they relate to the issue/problem described in Step 2 have a microbial problem, and that microbial problem needs balancing.

- Complete the SLG energy process work in the order needed (Step 6). Do not include the Microbial Balancing Program when determining the process order on your SLG Troubleshooting Chart.

- *Within 24 hours* after completing the energy processes testing positive on the SLG Chart, set up for testing the Microbial Balancing Program.

- Chapter 12 in the *MBP Manual* explains how to use the MBP Program for environments. You will now be using the MBP Troubleshooting Chart (not the SLG Chart) while working with the MBP Program for your project.*

- For Step 2 on the MBP Chart, write the name of the project and copy the description you wrote in Step 2 on the SLG Chart. This will ensure that the microbial work addresses whatever your soil-less garden needs as stated on that original SLG Chart.

- Test the MBP Chart and do the process work as indicated, using the language as written for the Microbial Balancing Program. Test which environmental coning (*MBP Manual,* Chapter 12) to use. Add the deva of your project to that new coning.

- Follow all the steps as written on the MBP Chart. Maintain all the timing for the processes and follow-up as written for the Microbial Balancing Program.

- You may continue your regular soil-less garden/project activity, including holding coning meetings, while you are working with the Microbial Balancing Program to address your project's microbial issue.

THE CHART

Troubleshooting Options

* If you have questions on how to apply this to your SLG, call our Question Hot Line.

(Continuation of the SLG Chart Steps)

Step 6. ORDER THE PROCESSES.

After you've tested the three Lists, test the order in which to do everything you checked in the Energy Process List, plus the Human List #1 and/or #2.

SLG TROUBLESHOOTING PROCESS OPTIONS ORDER

Let's say you are willing to work with all of the options listed under the Human Troubleshooting List (#3 through #5), plus the *Workbook II* Troubleshooting Process, plus the MBP Troubleshooting Process — and you've had the "sad luck" of more than one of these options testing positive. Here's your way out of what may look to you to be a major headache.

Test the order you are to do each of these options that tested positive. Then, *within 24 hours* after completing the SLG Troubleshooting Process and SLG Chart, begin working with whatever option tested to be done first. *Within 24 hours* after completing the first option and any needed follow-up, work with the option that tested second. And so on. This will spread out the optional work in a more manageable rhythm. Note that the options will set up their own rhythm and timing outside the work done with your original SLG Troubleshooting Process and your SLG coning. As long as you set up each option by reading/stating your project's name and what you wrote in Step 2 on your original Chart, you will maintain that SLG focus and intent when doing the optional processes.

> NOTE: If the options that tested positive are relatively simple, you may also complete them in one or two sessions, working with several back to back. Just maintain the order in which you are to do those optional processes.

Step 7. TAKE ONE DOSE OF ETS PLUS FOR HUMANS.

Take ETS Plus for Humans (10–12 drops) to restore personal balance that might be needed as a result of the SLG Troubleshooting work.

Step 8. TROUBLESHOOTING RECHECK DATE.

You may need to continue working with the focus of this chart in order to achieve full project balance. Ask:

> Is a troubleshooting recheck needed? (Test.)

If you get a negative response, continue on to Step 9. If you get a positive response, ask for the date. You may approach this by testing a calendar. Ask:

> What day should I schedule the recheck for this troubleshooting chart?

Then kinesiology test the calendar days. The day that tests positive is the date for the recheck. Note this date on the SLG Chart. (Then mark your calendar, add it to your PDA, put a big sign on your refrigerator, write it in lipstick on your bathroom mirror, teach your parrot to say the date every time you walk in the room . . . whatever you need to do to remind yourself you have an SLG Troubleshooting recheck to do.) You have *36 hours* from this date to do the recheck.

The point of the recheck is to allow for flexibility and the continuation of the Troubleshooting Process work in stages. Rechecks refine your SLG work.

TO DO A RECHECK:

- Get out the original chart you are rechecking *and* a clean chart.
- Open the SLG coning for the project you are working with. Take one dose of ETS Plus for Humans.
- Read Steps 2 and 3. This reestablishes the focus and intent of the testing.
- Test Step 4: Ask:

  Is a Troubleshooting Process recheck still needed for this issue/problem?

- If you get a negative response, the first round of work did the job after all and you are now clear on this chart.
- If you get a positive response, proceed with Steps 5 through 9 using a clean chart. NOTE: The results for Steps 5 and 6 may change in the recheck.*

MISSING THE 36-HOUR DEADLINE FOR A RECHECK:
Understand that the recheck work may be more complex and involve extra processes when it is being done outside the 36-hour rhythm. In short, if you miss the recheck date, you may have more work to do to bring the issue to completion, but you will not lose the benefits of the work already done. It is important to do a recheck, even if you are late. Here's what you do:

- Open the SLG coning and take a dose of ETS Plus for Humans.
- Read Steps 2 and 3 to reestablish focus and intent.
- Then test Step 4.
- If more work is needed, continue with the steps using a clean chart.

Step 9. CLOSE THE FOUR-POINT CONING FOR YOUR PROJECT.
Take one dose of ETS Plus for Humans (to restore personal balance that might be needed as a result of closing the coning).

* RECHECKS AND THE STEP 5 ESSENCES PROCESS:

- If the recheck includes another Essences Process solution, and you are still using an Essences Process solution as a result of the original testing, discontinue administering the original solution. Replace it with the new recheck Essences Process solution.
- If you test negative for a new solution, the old one is still effective and you are to continue administering the original dosage.
- Be sure to test for follow-up essences for any solution when the dosage is complete.

Chapter 5

Working with a Soil-less Garden and Its Rhythms

WORKING WITH A SOIL-LESS GARDEN isn't just working with the Troubleshooting Process. A project has an ebb and flow of activity that is constant. You may not be doing endless troubleshootings, but you will be addressing this ebb and flow, plus whatever your project needs at any given time. So there will be a lot of opening and closing of your SLG coning.* With soil-less gardens, you will be opening the coning and working with your team:

* See p. 21 for steps.

- For meetings that include others in the project who are working with you and the coning.
- For yourself alone when you attend meetings and the others present are not a regular part of the coning. Having the coning open will help you to maintain the project's decisions and movements in conjunction with the spirit of the DDP.
- Any time you want to bat around ideas with your coning team. Other human members of the team need not be present every time you open the coning. In this case, you would be having a private meeting with your coning team.
- Whenever you are actually working on a project such as writing a book, painting a picture, designing a brochure, acting in a play, delivering a speech, leading a class . . . These are examples of activity within the project. Doing them with the coning open adds a whole range of benefits. You're simply going to have to try this in order to see all it can give you.

● Once you get used to being in the coning, you may find it useful whenever working on an all-day project to open the coning in the morning, keep it open all day and then close it at night. (Don't forget to close it.) This will keep you in the constant ebb and flow of your soil-less garden and maintain any needed connection with your team. If you want to do this, read what I've written on Professional MAP *(MAP, Third Edition)* and working in day-long conings for professionals. I describe how to manage that coning situation and maintain your own balance within a long coning.

Some people think (hope) that if they set up a soil-less garden, their project will magically happen without their having to lift a finger. If this is your hope, don't bother stepping foot into a soil-less garden. With nature and the White Brotherhood working on your side, a lot of things can happen very quickly. If the pace is uncomfortable for you, open the coning and tell them you need for the pace to be a little "gentler and kinder."

Also, in a soil-less garden, you can never abdicate your role in the coning mix. You are the one in charge of making the day-to-day decisions about the definition, direction and purpose of your project. Definition, direction and purpose apply to more than just your project's DDP. They are a constant part of your project's flow. There will always be decisions for you to make, actions for you to choose and directions for you to go in. It's *your* project and no one else's. The buck, as they say, actually stops with you.

Neither nature nor the White Brotherhood will make decisions for you. However, they will provide options for you to choose from. Example: Your project has a situation you need to solve. You need more money coming in. So you take the situation to a coning meeting and explain it to your team as fully as possible.* You test to see if you need to do a Troubleshooting Process around the situation. You may not need one. Or you may need one. You do whatever is needed. In the meeting you did not pick up on any additional input about the situation from your team. Instead you tested that you were just to close the coning and get on with your day. This is akin to your team telling you, "No further discussion is needed and you may

* You are actually talking to your team. They are your partners and need to be kept in the loop with information from *your perspective.* The more you include them in the process, the more opportunity you give them to provide you with information from *their* perspective.

consider that your situation has been heard. Now go away." So a couple of days go by and suddenly three different options for funding fall into your lap. And the options came from three very different directions/people. In short, you never would have considered these options possible on your own.

Now you have three options staring at you. Some soil-less gardeners feel the thing to do at this point is to open the coning and ask which option they should choose. No. No. No. It's *your* responsibility to look at each option very carefully and then choose the one that appeals most to you or makes the best sense. *You* are the one in this mix who is responsible for supplying direction. It's *your* responsibility to decide which direction you wish your project to go in. Once you make the decision, the coning adjusts the pattern and flow within your project to reflect your decision. At that point, the insight and input from your coning team will take into consideration your decision. You don't have to ask them to make this adjustment. It will occur automatically once you make the decision.

When faced with multiple options, it is kosher to ask your coning one question: "Do each of these options fall within the guidelines and range of balance for my project?" Sometimes a well-meaning person who knows nothing about your project just feels they know exactly what you need and they'll toss some useless thing your way. It's quite possible that whatever they are tossing is actually something they personally need — and not you or your project. If an option falls in this category and you're not seeing it on your own, the coning will be happy to identify it for you when you ask this question. After that, it's up to you to make the decision to put that option aside and to choose which of the remaining options you prefer.

Being responsible for all the decisions around providing the daily definition, direction and purpose does not mean we don't need to actually get off our duffs and do something. A book won't be written by itself. (I've tried. It doesn't happen.) A painting won't paint itself. A class won't teach itself. In nature's definition of a garden, it states:

> *It* [a garden] *only needs to be an environment that is defined,*
> *initiated and appropriately maintained by humans.*

It must be appropriately maintained. In a vegetable garden, gardeners have to plant, fertilize, water, weed and harvest. In a soil-less garden, the gardener must provide all the physical and mental activity that is required for meeting the goals of the project and its DDP. Your coning partners give you the input for the best way to do this and the best way for doing it *in balance.* They give you insight on the best tools you need to work with and also provide you with the means for obtaining those tools, if only you'll look up and see what's right there within your reach. It's all there. All we have to do is recognize and respond.

Recognize and respond. An excellent phrase to remember. It moves us into the very heart of how we humans are to participate in a soil-less garden. We activate an SLG coning and garden and this sets off a lot of activity on all levels. Your coning team can never be accused of laziness. Moving the project forward often rests on our ability to recognize what's happening around us and respond to it — quickly. If you're going to the trouble of writing a DDP and creating a soil-less garden, your project's four-point coning members will consider that you are serious about achieving your goals. They'll take your project just as seriously.

Along these same lines of "job division," do all you can to resist the urge to take control of nature's role in the soil-less garden. Nature provides all the matter, means and action needed for your project. But there are those who just can't seem to help themselves, want to take control and try to do nature's job. Usually there's a trust issue here. No one else can do something better than this person, and they don't want nature "interfering" with all that annoying insight about matter, means and action. The truth is, we humans don't do a good job supplying matter, means and action *in balance.* We never have and we never will. We have proven that we don't know how to supply these things while maintaining the intricate dynamic of balance. If you need proof, just take a look at what we've done and continue to do to our global environment. We like to think we know what we're doing and only manage to further mess up the world's balance. The advantage of letting nature do its job is that the matter, means and action is provided in balance — with the project, with you and with the world around you. The bottom line: Soil-less gardens are environmentally friendly.

No matter how well-meaning we are, sometimes we can't resist the urge to take control. Usually it happens when something in the project occurs unexpectedly and things get a little tense. The knee-jerk reaction is to tighten our grip and take control. I suggest putting some thought into ways of recognizing when you do this. If you can "catch yourself in the act," you can respond either with a personal Calibration Process or by rededicating yourself to the coning process. It happens to the best of us, so don't waste time beating yourself up over it. The important thing is to *recognize and respond.*

About working with this team: Very few people actually hear their team members, unless it's their assistant, Harriet, sitting on the chair next to them eating chips. Other than this, there's nothing audible going on from the other coning members. To work with your team in meetings, use the setup laid out for kinesiology testing.* Ask a series of yes/no questions directed to your coning team. Then test for their answer. They'll project the "yes" or "no" into your electrical system, and it is this that you will be testing for. Tell Harriet to put down the chips and start taking notes. You need to note all the questions you ask and their answers. After your first meeting, you may think that you'll always remember everything that went on in the meeting. After the tenth meeting, you'll be so lost you won't know which end is up. So, keep notes.

* See Appendix A for steps.

Remember, your coning team gives you insight and suggestions from their perspective. It's a pretty lofty perspective and you couldn't ask for more expert input. However, they still won't be making decisions for you. It's your job to consider their input and then make the decisions. Over time, we all learn when we have made a mistake about a decision. But the coning members will not stop us from making the mistake. It's too valuable a learning opportunity for us. If we're serious about working in soil-less gardens, then we need to develop a good working relationship with our coning partners. We have an opportunity to learn a great deal about ourselves, especially in the areas of personal humility, listening, how we relate to others on a team and wise decision making. When we make a less-than-stellar decision, we have good partners we can turn to for insight into what made that decision weak.

Someone asked me in a recent Soil-less Garden Forum how to keep the information straight when they have several soil-less gardens up and running simultaneously. Whether you have one or more SLGs, set yourself up well. If you're computerizing your notes, create a folder for each soil-less garden. (This is not a bad idea since it gives you the ability to open a file and do a quick search for some specific information that you need.) If you are the pen/pencil type, get a binder and include a divider for each soil-less garden you have going. However, the real key for keeping everything straight is to keep good notes on all meetings, decisions, coning input, insights . . . Oh, yeah — and don't forget to date all your notes.

I think it's really important to choose as your first soil-less garden something that is simple. There is a lot for you to learn. You're learning about working in a coning team, what kinds of questions to ask and situations to bring up, how to do the simple logistics like keeping notes, who is going to keep everyone supplied with nuts (and chocolate) for all those meetings, and on and on. If you choose something simple — a small project like putting together and giving a fifteen-minute talk to your community club, or decorating one small room in your home, or organizing a closet, or making one presentation at work — you can keep the soil-less garden manageable while you go through the initial learning stages. Then you have a foundation for taking on other SLGs and, at the same time, you'll have a better idea of what a soil-less garden requires. This will help you to not over-extend yourself with some monster, but noble, project that is guaranteed to swamp you. And I especially recommend that you activate *one SLG at a time* when you are a beginner. Open one project. Complete it. Close it down once completed. Open a new project.

DEACTIVATING A SOIL-LESS GARDEN & ITS CONING

Some SLGs are long term and can go on throughout a person's life. In this case, the person's death automatically closes down that soil-less garden and the coning is deactivated. But most of the time, a project will be completed within a particular time frame and you will need to close everything down. You are the only one on the team who can deactivate a soil-less garden and its coning, just as you are the only one who could activate it in the first place. Closing it down is simple.

1. Open the project's SLG coning. Tell them that the project has met the goals of its DDP and has, therefore, been completed. You might want to spend some time having a little celebratory "office party" with your team. After all, you worked together on this project. When you are ready and the members of your team have started to dance around with lampshades on their heads, state:

> I would now like to close down this SLG and deactivate the coning.

Wait 15 seconds for the soil-less garden to gently close down and the coning to deactivate.

2. Take one dose of ETS Plus for Humans. You don't need to close the coning, because the coning is no longer active.

OPTION: INTENTS LISTS

I caused quite a stir during a recent Soil-less Garden Forum when I casually mentioned Intents Lists. I could have said that was a ringer I threw in just to see if they were paying attention, but it wasn't a ringer. It's something I've been working with for years and I talk about it in the soil-less garden video/DVD.

An Intents List is an "attachment" to a DDP. It is a list of attributes and qualities you want to see reflected in your project for color and flavor. I'm making it sound like a list of spices you might add to a recipe and, in a way, that's what it is. It isn't a DDP, but it modifies a DDP for a specific event or activity. Here's an example of

how I attach an Intents List to a DDP and what it is. The DDP for the Perelandra Open Houses that have been held on-site is:

- To introduce, interest and excite the visitors about Perelandra and co-creative science.
- To encourage, inspire and support anyone who wishes to know more about Perelandra and co-creative science. Give them confidence about making this move.
- To make those visitors who are not interested in Perelandra comfortable.
- To give all visitors the experience and feeling of Perelandra, what it stands for and what co-creative science can accomplish.

To this, I added an Intents List (a *really long* list put together by the staff) that set the tone and described the spirit of the Open House for all concerned (staff and visitors):

- Courteous/Considerate
- Thoughtfulness
- Ease
- Safety
- Fun/Enjoyment
- Sense of humor
- Calmness
- Efficiency/Effectiveness
- Appropriate education
- *Damn* profitable
- Groundedness
- Clarity
- Friendliness
- Teamwork
- Balance
- Flexibility
- Balanced impact on Perelandra
- Weather: Partly cloudy, low humidity, temperature in the upper 70s to low 80s Fahrenheit

We held open houses only three days per year, and I activated the Open House DDP a week prior to each date so that the final preparations would reflect the DDP. It helped keep the staff on track while preparing. When I activated the DDP, I also activated its Intents List. I deactivated the Open House DDP and List the day after each open house.

Our Open House Intents List is a lengthy one, but this is because of the unusual and intense nature of the day. The weather intent was something we hoped for but didn't always get. Actually, we did get a surprising number of open house days with comfortable weather. And then there were those other days. When we had tough weather (seriously hot and humid), I would look at the weather pattern covering several days before and after the open house. More often than not, we got the best weather of that pattern.

You don't have to activate an Intents List at the same time you activate a DDP. The Intents List may be "attached" to the DDP at any time. This is especially helpful when you have a special event or activity within an SLG project. You can add the seasonings that are relative to the event without changing the DDP for the entire project. Here's an example used by a soil-less gardener for the DDP covering his overall "Life Project":

> Intents List: safety, high return on effort and time invested, efficiency, cooperation, cleanliness, good hygiene, fun, "cool," innovative.

NOTE: From my Open House example you can see that a DDP written especially for an event may be activated at any time. The Open House DDP is not the same as the Perelandra DDP. It's a short-term DDP that falls under and is compatible with the Perelandra DDP. It's a DDP within a larger DDP. I set up this short-term DDP because the open house event, although clearly a part of Perelandra, is so unique in its timing, rhythm and activity. Having its own DDP added specific definition, direction and purpose needed for the event.

Activating & Deactivating Intents Lists

* SUPPLIES NEEDED: The *Companion* Activating & Deactivating Intents Lists steps, ETS Plus for SLG, ETS Plus for Soil, ETS Plus for Humans, pen/pencil, paper, spoon, paper towels, a watch or clock with a second hand.

ACTIVATING:*

1. Open the Project Coning.

2. OPTIONAL: Activate a secondary or short-term DDP using the Starting Process steps.

- If the secondary DDP has already been activated, read aloud that DDP to focus you and the coning on the DDP you wish to attach an Intents List to.
- If a secondary DDP is not activated and not needed, you will be attaching your Intents List to your primary project DDP.

3. State:

I want to set up for activating an Intents List with the following DDP:
Read the DDP aloud.
Wait 10 seconds.

4. Read aloud each attribute or quality on the Intents List, one at a time. Wait 5 seconds after each is read for its activation to occur.

5. Once everything on your Intents List has been activated, balance and stabilize the List and the DDP to which it is attached.

BALANCING:
For balancing, work with ETS Plus for Soil.

A. State:

I'd like to test if ETS Plus for Soil is needed for balancing this Intents List and the DDP to which it is attached.

B. Ask:

How many drops of ETS Plus for Soil are needed?

Do a sequential kinesiology test.** (See the Essences Process, p. 55.) Place the correct number of drops of ETS Plus for Soil in a clean spoon.

** If no drops are needed, move on to stabilizing.

C. You will release them to your project using an NS application.

To do this, state:

> I would like the essence of these drops to be shifted for any needed balancing for this Intents List and the DDP to which it is attached.

Hold the spoon out for 10 seconds while the shift occurs.

D. After 10 seconds, the shift will be complete. Record the number of drops you released. Wash the spoon. Do not try to save the remaining liquid because that ETS Plus cannot be used again.

STABILIZING:

For stabilizing, work with ETS Plus for Soil-less Gardens.

A. State:

> I'd like to test if ETS Plus for SLG is needed for stabilizing this Intents List and the DDP to which it is attached.

B. Ask:

> How many drops of ETS Plus for SLG are needed?

Do a sequential kinesiology test.* Place the correct number of drops of ETS Plus for SLG in a clean spoon.

* If no drops are needed, move on to the next step.

C. You will release them to your project using an NS application.

To do this, state:

> I would like the essence of these drops to be shifted for any needed stabilizing for this Intents List and the DDP to which it is attached.

Hold the spoon out for 10 seconds while the shift occurs.

D. After 10 seconds, the shift will be complete. Record the number of drops you released. Wash the spoon. Do not try to save the remaining liquid because that ETS Plus cannot be used again.

6. The Intents List and its DDP are now activated, balanced and stabilized. Close the coning and take one dose of ETS Plus for Humans.

* SUPPLIES NEEDED: The *Companion* Activating & Deactivating Intents Lists steps, ETS Plus for SLG, ETS Plus for Soil, ETS Plus for Humans, pen/pencil, paper, spoon, paper towels, a watch or clock with a second hand.

DEACTIVATING:*

1. Open the Project Coning. (If you have an Intents List attached to your main Project Coning and you wish for it to remain active throughout the life of the project, you do not need to do anything special to deactivate the List when you deactivate the project's SLG. The List is automatically included when you close down the soil-less garden.)

2. State:

   I'd like to deactivate the Intents List attached to the ________ DDP.

   Wait 10 seconds for the deactivation to complete.

3. If you want to deactivate a short-term or secondary DDP at this time, state:

   I'd also like to deactivate the following DDP. (Read aloud that DDP.)

   Wait 10 seconds for the deactivation to complete.

4. Balance and stabilize the remaining active DDP(s) and coning(s).

BALANCING:

For balancing, work with ETS Plus for Soil.

A. State:

   I'd like to test if ETS Plus for Soil is needed for the remaining active coning(s) and DDP(s) for balancing resulting from this deactivation.

B. Ask:

   How many drops of ETS Plus for Soil are needed?

   Do a sequential kinesiology test.** (See the Essences Process, p. 55.) Place the correct number of drops of ETS Plus for Soil in a clean spoon.

** If no drops are needed, move on to stabilizing.

C. You will release them to your project using an NS application. To do this, state:

   I would like the essence of these drops to be shifted to the remaining active coning(s) and DDP(s) for any needed balancing resulting from this deactivation.

   Hold the spoon out for 10 seconds while the shift occurs.

D. After 10 seconds, the shift will be complete. Record the number of drops you released. Wash the spoon. Do not try to save the remaining liquid because that ETS Plus cannot be used again.

STABILIZING:

For stabilizing, work with ETS Plus for Soil-less Gardens.

A. State:

I'd like to test if ETS Plus for SLG is needed for the remaining active coning(s) and DDP(s) for stabilizing resulting from this deactivation.

B. Ask:

How many drops of ETS Plus for SLG are needed?

Do a sequential kinesiology test.* (See the Essences Process, p. 55.) Place the correct number of drops of ETS Plus for SLG in a clean spoon.

* If no drops are needed, move on to the next step.

C. You will release them to your project using an NS application.

To do this, state:

I would like the essence of these drops to be shifted to the remaining active coning(s) and DDP(s) for any needed stabilizing resulting from this deactivation.

Hold the spoon out for 10 seconds while the shift occurs.

D. After 10 seconds, the shift will be complete. Record the number of drops you released. Wash the spoon. Do not try to save the remaining liquid because that ETS Plus cannot be used again.

5. This deactivation is now complete. Close the coning and take one dose of ETS Plus for Humans.

Chapter 6

SLG Questions & Comments from the Perelandra Forums

THE FOLLOWING QUESTIONS and comments were recently submitted during our Virtual Open House and Soil-less Garden Forums. Even if you have already read the 2007 Forums posted on our web site, you'll probably find it worthwhile to go through the questions included here again. I took the opportunity to smooth out the wording and flesh out my answers.

> NOTE: When reading the 2007 Forums, remember that I streamlined the processes in the *Companion* by weaving in ETS/Soil-less, ETS/Soil and ETS/Humans *after* we held the Forums. The answers I gave to questions about using essences and the Soil Balancing Kit became outdated when I wrote the *Companion*. If you have steps from different Perelandra material with differing instructions, use the information I've given you in the *Companion*. The *Companion* contains the latest and most up-to-date information.

DDPs

A LIFE'S DDP

Question: I don't particularly like my job, but doing soil-less garden work with it seems to make it more tolerable. One of my reasons for keeping the job, besides the money, is that I've assumed that (1) it's necessary to be flexible, and (2) no matter how much I don't like the job, I should be able to make it enjoyable by working with nature through soil-less garden work on it. Are these assumptions correct?

Machaelle's Answer: The problem with your question here is that I'm not the one to answer it for you. Only you can answer it. You are in charge of your own DDP. If you feel you have to stay in this job, and you're working with nature in an SLG coning regarding that decision, you'll get from nature the matter, means and action to make staying in the job palatable. But it's your decision and your decision alone on whether you want to stay in the job or if it's right for you to stay in the job. Whatever your decision, your coning partners will support that.

DDPs

Question: I find it challenging to create a DDP that will be in balance with the whole of life. I mean, how does one really *know* in the beginning what the ramifications will be of creating that DDP? What guidelines do you suggest in getting clear with nature that what you are setting up will be for the good of the whole?

Machaelle's Answer: Okay. You figured out that writing a DDP can be challenging! Here's the point of the partnership: You define, direct and give purpose to what you want to do, and nature will take that information and provide the matter, means and action in balance. *Nature automatically balances it "for the good of the whole."*

If you end up with something other than what you had wanted for your project, you probably need to go back and rewrite or amend your DDP. And if you forget your role in this, and also try to define the matter, means and action for achieving your DDP, you probably will experience imbalance in your project. So there are two things that you're looking at:

- *Stay out of nature's way in the partnership. Don't do nature's job.*
- *Understand that your development as a person with free will is constantly challenged and questioned according to whatever it is you're asking for.*

You'll learn in time that sometimes what you're asking for is inappropriate. Your coning partners are going to reflect back to you exactly what you ask for. It's your job to look at it and decide if you think what you ask for is responsible. So when you're new to soil-less gardens, start small so that you can start learning about DDPs and what it takes to create a project that includes balance.

SLG CONINGS

SETTING UP CONINGS

Question: As a newbie to this whole co-creative process, I get confused about which devas and which WB department members to have in the coning for soil-less gardening. Is it OK to let the team decide once I clarify what I am working on, or do I need to specify?

Machaelle's Answer: See Chapter 3: The Starting Process, step 3B.

SLG: BUDGET / FINANCES

Question: In setting up a coning for a project, let's say my budget, I would include the Deva of Budgets, but should I also include the Deva of My Finances? I worry the Deva of Budgets only knows about Budgets but the Deva of My Finances really knows the scoop about my financial situation.

Machaelle's Answer: I would only include the "Deva of My Finances" in the coning because finances include budget. Budget is a function of finance.

THE SLG CONING SETUP

Question: I noticed that there is reference to including devas (plural) in a soil-less garden. I might be due for another viewing of the SLG video because I thought that once you establish the DDP and start the garden with the initial calibration that you work with the deva of that garden when opening the coning. Please clarify: Do I include the deva of that SLG or the devas for the SLG or particular issue(s)?

Machaelle's Answer: Most of the time and in most SLGs you are working with one deva only — the deva of that project. But in more complex projects such as a business that includes different departments, it may be necessary for you to add the deva of a specific department in the coning when addressing issues from that department. For example: Most of the time I have just the Deva of Perelandra activated in the Perelandra SLG coning. However, when dealing with a specific issue in the Production Department, I have two devas in the SLG coning: the Deva of Perelandra and the Deva of the Perelandra Production Department. This focuses the coning temporarily on one specific area of Perelandra. When I'm finished with the production issue or meeting, I disconnect the Deva of the Perelandra Production Department from the main SLG Perelandra coning.

To include a departmental or additional deva to the Project Coning for meetings or an SLG session: .

Open the Project Coning and say that you would also like to include the Deva of ________. Wait 10 seconds.

Check to see if the expanded Project Coning needs to be balanced and stabilized. (See the steps for balancing and stabilizing, p. 50.) Frequently, this will test clear, no balancing/stabilizing needed. Finally, take one dose of ETS Plus for Humans.

When the meeting is over and you wish to keep the Project Coning open, release the deva of the department by stating that you would like that deva to be released from the Project Coning. Wait 10 seconds for the release. You now have just the Project Coning open. If you want to close the expanded coning that includes an additional deva, just follow the steps for closing an SLG Project Coning. Be sure to take another dose of ETS Plus for Humans once this work is complete.

CONINGS WITH ADDITIONAL PEOPLE

MULTIPLE PEOPLE IN THE CONING

Question: Can you do a soil-less garden in tandem with another person — with their agreement and interest and DDP? (family member, client, friend, pet)

Machaelle's Answer: Yes, but to keep the arguments down, only one person should be responsible for kinesiology testing any questions that need to be asked.

Since a soil-less garden is defined, initiated and maintained by humans, then those working in tandem with you should also be human. The nature element of the soil-less garden coning covers the information that you feel an animal or pet might provide.

CONING SESSIONS WITH OTHERS IN THE ROOM

Question: Would you please give some information on energy processes with an activated coning while others are present in the room? For example, when I do an Energy Cleansing Process for my SLG with others present, how/when do I include their higher selves when opening and closing the conings? And do I verbalize aloud what I am visualizing (i.e., the sheet moving up) during the process, so all can focus, or do it silently and let them just be energetic support?

Machaelle's Answer: When working while others are present and wish to be supportive, just open the coning and include only your higher self. Then go through the process, saying the steps out loud so everyone can follow along with you. One nice thing you can do after closing the coning is ask if any of them got any insights or pictures or words during the process that they would like to share.

CONING SESSIONS WITH MULTIPLE PEOPLE

Question: 1. Can I, with permission from my SLG partner who is unavailable at the time, have a coning meeting and include his/her higher self in the coning?

2. What is the best setup for a business garden with multiple divisions? Will each division have its own deva and WB members? What's the wording you use to open a division coning?

Machaelle's Answer: 1. No. Only include the higher self of the SLG team members who are physically present at the time.

2. I can't tell you the best setup for your business. You have to ask your Project Coning if something special needs to be set up for different divisions. Sometimes the answer will be yes, sometimes the answer is no. If you are to adjust the Project Coning for your company departments, simply add the "Deva of ________" (department) to the regular four-point Project Coning for that department.

CONING SESSIONS WITH PEOPLE IN AND PEOPLE OUT

Question: 1. I am working with a committee of nine to bring a Symposium to town. Three of us understand the project as a soil-less garden, and we open conings to state the DDP, balance the project with ETS, report in, ask questions, etc. Is this effective/kosher to do without the other six committee members included in the SLG?

2. When all three of us are not available, I will open a coning to balance, etc. At this time, do I include the higher selves of the other two who are involved but not present?

Machaelle's Answer: 1. In this situation, it is only appropriate to set up an SLG to cover the roles for the three of you within the Symposium and with the larger committee. It would be easier if each of you set up your own personal soil-less garden with DDPs that cover your individual roles within the Symposium and committee.

To set up an SLG with just three of you willing to recognize it as an SLG puts the other six people in an awkward position. This is the part that is clearly not kosher. You rectify this by focusing your individual SLGs solely on how you do your particular job for the Symposium and meet your responsibilities in the project.

2. Never include the higher self of someone who is not physically present and consciously participating in the coning.

CONING SESSIONS & WORKING WITH OTHERS

Question: When working together with a friend on a co-creative project, how do we organize the conings? Can one of us open a coning with both our higher selves to work on an issue? (Yes, before you hit me, we are both present and agree the higher self should be included). If one person needs to leave, can we just disconnect that person's higher self and the other can continue to work? What if the person leaving is the person who opened the coning at the beginning? Can the other person keep the coning open to work and close the coning? Basically, what are the do's and don'ts with conings when we work on a project with another person who is also into co-creative science?

Machaelle's Answer: Here's the multiple-person soil-less garden protocol: One of you takes the lead, opens the coning and includes the other person's higher self in the coning. If one of you needs to leave before you're finished, that person disconnects himself from the coning. (Just request that your higher self be disconnected from the coning and take one dose of ETS Plus for Humans. Then leave.)The coning is still open with the remaining person's higher self, and that person can continue working. It doesn't matter which person opened the coning. As long as the coning population reflects the people whose butts are in the room at the time. (NOTE: In an SLG with multiple people, the person who opens the coning is simply the spokesperson and not the sole owner of the coning.) Whoever is left in the room then closes the coning.

MULTIFACETED BUSINESSES

QUESTIONS & COMMENTS

A PROJECT WITH DEPARTMENTS

Question: When developing and presenting workshops, is each workshop a separate SLG even if presented at the same conference? Can I have a general SLG for "professional presentations"? Do I include the devas of the conference or just for my workshop garden? Or do I ask to be connected with "appropriate" devas?

Machaelle's Answer: Here's one way to approach this situation: If your work (your job) is to develop and present workshops, that is the focus of your soil-less garden. You will have a project DDP and coning for this. For each workshop, open your main soil-less garden Project Coning and activate an Intents List. (You may include qualities to address the individuality of a particular workshop.) The Lists remain activated throughout the workshop or conference, even if the event includes more than one presentation. Just before each presentation, spend a minute or two focusing on that presentation's Intents List. This will connect you with the correct List at the correct time. After the event is concluded, reopen the Project Coning and deactivate each of the Intents Lists. Then close the coning.

AN SLG WITH DIFFERENT DEPARTMENTS

Question: I have a soil-less garden for my work. The DDP includes counseling, writing, teaching, recordings and ceremonies. When I am counseling, I work with a Professional MAP team coning. Is it wise/appropriate to also call in the Deva of my SLG, or the Deva of Counseling, when counseling?

Machaelle's Answer: Work with Professional MAP when you are actually working with any health practice or getting information about a health practice. Work in the soil-less garden coning for your writing, teaching, recordings and ceremonies, plus when addressing the supportive structure of those activities. Just to be clear — when you are counseling, you are working with your Professional MAP team. (The MAP coning includes the Deva of Healing and that covers counseling.)

AN SLG WITH SUB-PROJECTS

Question: When doing a soil-less garden with my foundation DDP, and within the foundation DDP I have several other sub-projects going on at the same time, is it appropriate/a good idea to complete DDPs for being more focused with each sub-project?

Machaelle's Answer: It's always helpful to have that kind of clarity, even when you're working with sub-projects. You can even activate a "sub-project DDP." But, you need to address the issue with your Project Coning first. Ask if the sub-projects need to be treated as separate divisions with their own amended conings (i.e., the main coning + the deva of that sub-project or department). Then ask if any of them would benefit from a DDP that is more directly responsive to the sub-project's focus. If you get a "yes" to either question, don't assume it means all your sub-projects need to be treated as separate divisions or that they all need their own DDPs. Test a list of your sub-projects to find out which need to be treated as a department with their own modified coning. Then test the list to find out if any would benefit from an individualized DDP.

Another question to ask: Are any of these sub-projects actually their own soil-less garden that would be best worked as a project that is completely independent from this original soil-less garden/project? If you get a positive response, test the list to find out which ones need to be set up as independent soil-less gardens. For this, you'll need to write a new DDP for each and put each through the Starting Process.

The following example may be helpful for understanding a coning/DDP breakdown for a business with five different departments.

Main SLG Project DDP and coning: Your overall business and its foundation four-point Project Coning

Department 1: the main Project Coning + the deva of this department

Department 2: the main Project Coning + the deva of this department

Department 3: the main Project Coning + the deva of this department.

Add a departmental DDP: This department would benefit from having its own DDP that is compatible with the main project DDP and reflects the unique elements of the department. Write this DDP and activate it using the Starting Process.

Department 4: the main Project Coning + the deva of this department.

Department 4's special project (i.e., putting together a fall advertising campaign):

The main Project Coning + the deva of the department + the deva of this special project. Include a DDP for the special project that is also compatible with the main project's DDP.

Department 5: the main Project Coning + the deva of this department.

- *When you are working with the SLG principles for departments 1, 2 and 5, you simply open the main Project Coning, add to it the deva of that department and have your SLG session as usual.*
- *When working with department 3, you'll need to write a departmental DDP and activate it using the Starting Process. Once this is done, the soil-less garden work in that department is done within the main Project Coning + the deva of the department.*
- *When you are having a regular SLG meeting in Department 4, open the main Project Coning and add to it the deva of that department. However, for the fall advertising campaign, you will need to write a new DDP that is focused on the goals of the advertising campaign. Activate the DDP using the Starting Process. Then conduct all the SLG meetings and work on the campaign within an open main Project Coning + the deva of the department + the deva of the special project (the fall advertising campaign). Once the campaign work is complete, deactivate its DDP. You will then go back to holding departmental meetings and working with the main Project Coning + the deva of the department. For the next special project in this department, ask if you need to write and activate a new DDP for it. If you test "yes," add the deva of that new project to the regular departmental coning.*

A COMPLEX BUSINESS SETUP

Question: I have a small health care practice testing others for essences and herbs. I would like to work with Professional MAP, but I am confused about which parts of my practice/business belong in Professional MAP and which are a soil-less garden.

Machaelle's Answer: Professional MAP works with the therapy you are providing and any tools needed to provide that therapy. For example, a massage therapist would include in Professional MAP questions about the oils that are used, the kind of table that's used, the technique of the massage as it relates to individual clients, the therapist's personal ability to do massage, etc. But if it's the elements that are required to create and maintain a business such as money flow, accounting, taxes, files, computers, staff, phone systems, the location of the office, advertising . . . that's the soil-less garden.

If you have a question that falls into a gray area and you don't know which category something falls under, open either coning (Professional MAP or your Project Coning) and ask which coning should be used to address the area in question.

WORKING WITH MULTIPLE SLGs

ORGANIZING MULTIPLE SLGs

Question: I worked with a soil-less garden (my counseling practice) for a number of years. Out of that I saw the benefit of working with nature in every garden. Sometimes, I get ahead of myself by thinking about the big all-life gardens instead of working with the processes to clarify and act. How do you keep track of the pieces?

Machaelle's Answer: First, don't think so big! But if you have several soil-less gardens in mind that are practical to your current life, get a notebook and dividers. Each soil-less garden has its own section. Everything you ask, all the information you get and all the testing you do is recorded and dated in the notebook. Of course you could also set up all your SLGs on your computer. Each SLG has its own folder.

MULTIPLE SLGs

Question: Can I activate and work with a second garden as things get going?

Machaelle's Answer: Yes, you can work with more than one garden at a time. But you need to keep your wits about you and only activate as many gardens as you are comfortable managing. That includes maintaining each one properly. You need to be realistic about the time and energy you can devote to several SLGs. I suggest starting with only one relatively easy SLG just to go through the learning curve and to get an idea of how much time and effort is involved. Do this before you open multiple SLGs.

MULTIPLE SLGs / MONEY GOALS

Question: My wife and I have set up a DDP together for our combined efforts to create a life in the country untethered to the financial lifeline of city life. We have joint meetings to discuss finances, planning, etc. We each have our separate DDPs for our individual projects, and work on these separately. Is that appropriate? Can I open a coning by myself for the combined project (she often does not have the time) to do the energy processes? I suspect we are controlling the financial part of this by stating our money goals. Is this ever appropriate?

Machaelle's Answer: 1. If these separate projects are related to the combined efforts to create a life in the country, then you need to find out from the Project Coning whether you need a separate DDP for the individual projects and, if so, which ones. For some of this activity, you may have a more direct relationship to the main DDP than you realize and you can save yourself some steps by working under one "umbrella" DDP and project.

2. Yes, you can open a coning by yourself for the combined project to do the energy processes.

3. You have to be careful how you're stating your money goals. You've already stated your money goals with the phrase, "a life in the country untethered to the financial lifeline of city life." However, because of how you're approaching this, you may end up churning your own butter and making your own furniture in order to achieve your DDP's goals. So as part of your DDP, describe the lifestyle you wish to achieve. But be general in this description. For example: a lifestyle in the country that includes solar power, electricity, an indoor toilet and running water.

CHANGING A BUSINESS FOCUS & EXPANDING ITS ACTIVITY

A PROJECT'S NEW ACTIVITY

Question: I've been writing a novel with the assistance of a coning group. Now I'm moving into the marketing phase. Do I continue working with the same group? Form a new group? Ask that the existing group reshuffle?

Machaelle's Answer: It depends on whether or not your original DDP included the activity around the marketing process. If it did not, open your original Project Coning and ask if they recommend that you simply amend the original DDP to include marketing as a department of the original project. Or would it be better to activate a new SLG with its own DDP and coning that are focused on marketing? Let your coning partners help you with this situation.

STARTING A NEW BUSINESS FROM AN OLD BUSINESS

Question: I recently changed my business from a print magazine to going online exclusively. In some ways, the online business is just an evolution and expansion of the old, but one could also say that the online business is completely new, though certainly built on the connections, experience, cache of content and office equipment of the old. Any hints on how to shift the energy to the new and write the DDP appropriately? Do I put the old one "to bed" after "harvesting" it?

Machaelle's Answer: Even with the connections to the print magazine, your online magazine is a new business with its own rhythms, patterns and needs. My suggestion is to treat it as a brand new business. The easiest way to do this is to deactivate the DDP for your old business and immediately turn around and activate a new DDP that addresses your new business. That will keep everything clean. Putting together a new DDP will be helpful in clarifying your new direction with the online magazine. If you have any remaining links that are important to the old business, they will come up as you're putting the new DDP together and they can be part of the new DDP.

IS THIS A "KOSHER" SLG?

"STUPID" IDEAS

Question: Rank beginner here . . . sorry . . . my first question: Will nature come right out and tell me my project is a stupid idea? ("Hey dummy, the division is going west, the company east, the marketplace south and you're trying to head north. Get a clue.") Maybe some SLGs are just bad ideas. How do I find out without muddying the water beforehand?

Machaelle's Answer: OK, that gave me a good laugh. Try this: Write out your DDP, activate it by putting it through the Starting Process. Wait the 24 hours after activating. Then direct the following statement to your Project Coning: "The first thing I want to know is if this project is a stupid idea and should be scrapped right now?" Tell your partners that you are open to getting information about this within the next seven days. Close the coning and return to your normal life. For the next seven days, notice the information that points towards the answer to your question, even if the information seems contradictory. These pieces of information will give you the input you'll need for deciding if this is a stupid idea or not. If you decide it's stupid, thank your team for its help, deactivate that SLG and try again.

Remember: You are the one in this equation who is responsible for deciding if something is appropriate or not. But there is no SLG rule that says you can't get help from your project members so that you can clearly see the reality of your project and how it fits into the bigger picture. By addressing the Project Coning with this, you are not asking your partners to make the decision for you. If you did, they would simply say, "Sure, you can do this." They will not make a decision on whether your SLG is a stupid idea or not. That's your job.

IS THE HUMAN BODY AN SLG?

Question: I am new to this soil-less garden thing and I am having a bit of trouble understanding it. My question: Is our body considered a soil-less garden or does that fall under another description? Sorry if this is a silly question.

Machaelle's Answer: This isn't a silly question — particularly when you're new at this. The human body is a garden. But at Perelandra, we have a separate collection of tools for addressing the health of the human body: MAP, the Perelandra Essences, the MBP Balancing Solutions, ETS Plus for Humans and the Microbial Balancing Program. You don't have to make a soil-less garden out of it. But if you have a general health-related goal, you can make an SLG out of that. See p.14 for two examples. Or you can work with your MAP team for achieving these types of health-related goals.

SLGs AND RELATIONSHIPS

Question: I've been working in an SLG to manifest a relationship. It's been an amazing process and I'm learning a lot, but I have a few questions:

(1) Is an SLG just a more clear and conscious way to work with the first dynamic of manifestation that you describe in *Behaving* and *Co-Creative Science?*

(2) Is it "kosher" to work in an SLG to bring about a desired relationship? I don't see many questions about relationships but it seems like they fit the criteria.

(3) My team often tells me to relax and let them handle the timing and the context (matter, means, action, right?), and they also put me to sleep often so they can work on deeper levels than I'm conscious of. Based on others' answers I'm going to do a calibration for these timing and trust issues, but it would be nice to get your input on whether these things seem normal or common in your SLG experience.

Machaelle's Answer: 1. An SLG sets up the framework for a more conscious participation with nature than the simple dynamic of manifestation. And in this age of environmental disaster, it's really good to promote the conscious relationship with nature so that we can stop making environmentally unfriendly decisions.

2. First of all, relationships do fit the criteria. So, yes, it is kosher. However, the best focus for that kind of soil-less garden is to prepare yourself for a relationship so that you will draw to yourself the right relationship in the right timing. The main thing is that you need to be prepared.

3. You can ask your Project Coning if it would be more beneficial for you to take these personal issues to MAP. Everybody's experience with soil-less gardens is different, but I think you fall within the range of "normal and common."

USING AN SLG TO PREDICT THE FUTURE

Question: What do you do when the responses you've gotten are seemingly wrong. I mean you are told something will happen, or is happening, and it doesn't?

Machaelle's Answer: You're asking the wrong questions. You're trying to get a prediction. Predictions only work when six billion people do and think precisely what they were doing and thinking at the exact moment you asked for the prediction. Obviously, that's not going to happen. You need to concentrate your questions on what you can act on now and let the results fall in place in their right timing without you trying to manipulate it. I'm suspecting there's fear around trusting the process and you may need a personal calibration around this.

IT'S NOT AN SLG

Question: One of the components of my soil-less garden is "my gardening work." Should this component really be "my garden" and not a soil-less garden? Is it helpful to have a soil-less garden called "my gardening work" and a non-soil-less garden called my garden, or do I really only need "my garden" as a non-soil-less garden?

Machaelle's Answer: Have you been drinking?! You say something like this to nature and I bet you'll get a great big, "Huh?" back. You do not need a soil-less garden for your plant gardening work. Just have a garden and a simple garden coning and enjoy.

AN SLG ELEMENTAL ANNEX

Question: Would you please give some examples of an Elemental Annex for a soil-less garden? I have used a potted plant in my office, and included the surrounding area. I wonder, however, is part of the business itself to be designated as such? Any other examples you could offer?

Machaelle's Answer: For a soil-less garden, you don't have to use a live plant for your elemental annex. An Elemental Annex is a gesture, a gift, from you to nature. So, it can be a picture on a wall (that doesn't need to be watered or dusted). Or a rock. Or a small handcrafted thing. Any small thing you would like to offer nature intelligence as a gesture of recognition and appreciation.

MORE ON THE ELEMENTAL ANNEX

Question: Elemental Annex for a soil-less garden! That's news to me. I don't remember reading that one was required. Are they required?

Machaelle's Answer: No, they're not required. This is something that is discussed in Workbook I *(Chapter 3) and people are applying that gesture to their soil-less gardens.*

HELPFUL TOOLS FOR THE SOIL-LESS GARDENER

NATURE CARDS

Question: So, you've done the DDP, you're going along, getting responses to your questions, feeling like you're on the right track, and everything goes awry, or feels like you're moving through mud. Any suggestions?

Machaelle's Answer: Yes. The Nature Cards. At these times, what you need is a hint/clue for getting unstuck. In this situation, you've lost sight of the project's rhythm and direction. The other thing you can do is open the Project Coning and tell them exactly what you've written here. Then say, "I need some hints to get out of this." Close the coning, get on with your day and expect the hints to come to you as you move through the next few days. Someone will say something, you'll get an idea or insight, you'll read a magazine article or see something on television . . . Be alert and expect the input to come from both ordinary and unlikely places.

NATURE CARDS

Question: Calibration is where I get stuck. I don't get any insight and, therefore, no resolution. So I never know what to do next, I become frustrated and, sadly, I abandon the soil-less project.

Machaelle's Answer: Try the Nature Cards. You're getting in your own way. Let the coning shift information about the calibration to you through the cards. This is why the cards were developed.

VOICES

Comment: I had a very difficult time figuring out the SLG processes in reality. I got it in theory right away. I ordered all the past issues of *Voices* (the Perelandra newsletter) and read them over and over again. Learning how others used SLGs, their success and their difficulties certainly helped me on my merry way with SLGs. It's an endless lesson plan and, as with everything co-creative, the "garden" getting the most benefit is the gardener.

ETS PLUS FOR HUMANS CLEARS UP MENTAL MUDDLE

Comment: When I get muddled and feel I am not communicating effectively with nature and the other Project Coning partners, I take ETS Plus for Humans and the muddle goes away. Sometimes it takes a few doses but it usually works.

MAP / CALIBRATION

Question: For a person to be in a place of greater clarity and develop more for personal evolution, would this not be a role within MAP/Calibration for any mental and emotional needs or concerns, along with an SLG project for personal evolution? Isn't this a place where nature can help with the whole process?

Machaelle's Answer: Yes, you're right. If you're struggling to gain clarity, taking the issue to MAP/Calibration is a wise thing to do.

THE ENERGY CLEANSING PROCESS IN BUSINESS

Question: This is regarding the Energy Cleansing Process and other energy processes for a business location. It is not convenient for the business to have me do an Energy Cleansing Process during business hours, and it's not convenient for me to work outside of business hours (before 6 A.M. or after 8 P.M.). What are the guidelines for working remotely? Can I go there, open a coning and go elsewhere to do the actual work?

Machaelle's Answer: As long as you have permission to do this work for the business, you can do the Energy Cleansing Process from home. And by using the new labeled circles to represent the business you are cleansing, along with the steps as written in the Workbook II, *this work is even easier. Even though you are doing this work from home, choose times when nobody is in the building. Some people can feel the Energy Cleansing sheet go through them and it freaks them out a bit (even though it's perfectly safe).*

USING ETS PLUS FOR SOIL-LESS GARDENS

Comment: It was noted in the Soil-less Garden Forum that people should write in how they worked with soil-less gardens, and I wanted to share with others how I've specifically worked with the ETS Plus for Soil-less Gardens.

When it was made clear to me through some of the basic definitions that nature has provided Machaelle — that nature is the essence, intelligence and structure of form itself; and that whenever a human being has a direction, definition and purpose for something, nature is automatically involved — when all of this was made clear, I realized that my relationship with anyone or anything that I impact or that I am impacted by, is a garden, of sorts.

So, lately, I've taken to thinking about my relationships with people, organizations and ideas that affect me significantly as perpetual projects which could use (often) a dose of stabilization and balancing. To that end, I've been using ETS Plus for Soil-less Gardens to enhance the order, organization and life vitality of said relationships.

I do *not* try to affect the things/people/ideas themselves, merely my relationship to them. And I can say that each thing that I apply ETS Plus for Soil-less Gardens to has gone from being a hazy reality of ambiguity or stress or even unpleasantness to an almost instant feeling of strength and stability.

In these situations, I merely hold out eleven drops of ETS Plus and say, "I ask that these drops be released to the order, organization and life vitality of my relationship with _________."

Usually I feel the essence going out and then, about a minute to five minutes later, I feel an echo coming back and usually a significant shift in how that person/issue/organization/idea makes me feel and how I feel about them/it.

Anyway, I just thought I'd share that little technique with those who use ETS Plus and the Perelandra processes. Wherever there are thoughts, feelings, attitudes and intentions, there is form and, therefore, nature.

TROUBLESHOOTING A COMPUTER NETWORK PROBLEM

Comment: We were having trouble with our computer network at work. First we couldn't access the Internet, then we couldn't access our server or send e-mail, then some computers could access the Internet but not the server, then . . . Well, you get the idea. We worked on it for two days and had finally narrowed the problem down to some wiring between the building with the Internet access and the building with the server. But we kept coming up against obstacles and complex "fixes" that could take days or weeks, and cost thousands.

Finally, after doing a routine Energy Cleansing Process for the business, I thought to do a troubleshooting on the network problem. I had the regular business coning open, so I was already with the right team to ask. (I think they may have given me the idea to ask, actually.)

The first troubleshooting test was for "Server function on internal network and Internet (external) general balance." It needed eleven drops of ETS Plus for Soil-less Gardens. Then I did a second test for the specific problem of the "connection between buildings isn't working." It needed six drops of ETS Plus for Soil-less Gardens. No other testing or troubleshooting processes were needed for either focus.

I closed the coning and went back to work. About fifteen minutes later I found out that our fiber optic guy had called and had given instructions to another staff member on how to rework the wiring with the fiber optic cable we already had in place. Within an hour and a half, our network was back up and running.

KINESIOLOGY

Question: I know that *practice* makes perfect when working with kinesiology and I am often confident about the results I get. But sometimes I feel that I'm not close to being on target with it [kinesiology] working correctly. Does my attitude (confidence versus lack of confidence) play a role in the testing accuracy?

Machaelle's Answer: Yes. Your attitude can throw a kinesiology test. Fear, lack of confidence, desire to control, doubt . . . And it often causes you to double-check and triple-check yourself, which can really lead you into a mess. The only way to get through this when you're feeling a lack of confidence is to take a moment, buck yourself up, commit to getting "the" correct answer no matter what it is and just go for it. Do the test and consider that this is the correct test result. Don't retest. Act on your first result. More often than not, you'll see that your testing actually was correct. This is how you restore your confidence.

POTPOURRI

THE SLG TRIANGULATION PROCESS

Question: What do you think about using the Triangulation Process to balance multiple soil-less garden projects? Could a soil garden also be one of the points? It seems to me that it shouldn't matter which project team one works out of because all points/lines will get balanced.

Machaelle's Answer: Whatever you test is needed, that's what you do! You'll have to ask the Project Coning for the direction on how to set this up, including if a soil garden may be one of the points. The important thing is that you set up a Triangulation Process according to what your coning partners are telling you.

Also, the Triangulation Process needs to be done from the perspective of whatever Project Coning you're working in, no matter which points and links you are working with. Doing this process for one project doesn't mean that you've automatically balanced the other projects from their own perspective. You'll have to test within each Project Coning. Finally, in the SLG Troubleshooting Process, you do not have to identify the points of the triangle. The Project Coning does this for you.

PETS IN SLG MEETINGS

Question: Is it a problem to have pets hanging around for the SLG meetings?

Machaelle's Answer: A pet can be in the room. But if you notice your pet is getting restless, the coning energy may be bothering him and he may need to be moved to another room. Other than that, pets hanging around are fine . . . unless they fart.

FROM THE SLG PHILOSOPHY DEPARTMENT

SLGs AND A PERSON'S FREE WILL

Question: When I use the ETS Plus Solutions, impossible-to-schedule people appear at events when I need them. Obviously they retain their free will. Do you think the ETS Plus balancing impacts their free will by changing the environment so that they make a choice that is aligned with the project?

Machaelle's Answer: Interesting question. The SLG balancing is not impacting their free will. It's changing the environment so that they feel drawn to the event and feel comfortable to schedule themselves to get there on time. It's assisting their ability to arrive at an event and not manipulating or forcing them to arrive at the event.

COMMITTING TO THE TRUTH

Question: In your Soil-less Garden video, you tell of the woman who does not water the tree because she is hot and tired and her kinesiology reflects this. She asks the tree if it needs water and the answer is "no" when it really does need water. I am afraid that when I ask nature where to put my money in my budget, the answers I get with the kinesiology testing might be based on my desires. I am not always sure the right answer is getting through. When you say "evolution," do you mean that my purpose is to do what I believe nature is telling me to do?

Machaelle's Answer: Evolution includes committing yourself to truth, no matter what your personal desires are. In the case of this example, a commitment to the truth would have given her the test result that indicated that the plants needed to be watered. Instead, her resistance to do this job on a very hot, humid, sunny afternoon overrode her testing. This is a way we can throw a kinesiology test. Had she recognized her resistance and committed to getting the truthful answer, she probably would have also gotten the information on how to complete this task in a comfortable manner. (Like asking someone to help her.) But she had already decided that this was going to be a tough job, so she sabotaged the test. When we ask nature and our other coning partners for information, it seems to me it's a useless exercise unless we're willing to accept the answer. If not, we can skip the SLG thing and just have a conversation with ourselves. Perhaps the thing for you to do is to put aside your soil-less garden on budget and finance and activate a more simple soil-less garden so that you can work with nature and practice the give-and-take that goes on with your coning partners. Pick something that has fewer "emotional land mines" than your current financial SLG. This will result in gaining greater confidence when you go back to your financial SLG. Working with finance and money can be quite an emotional exercise. So learn how to recognize your emotions and how to put them aside in order to get accurate information before you step into a "mine field" like finance and budget.

FOCUS ON THE EVOLUTIONARY SIDE OF THE I/E EQUATION

Question: I'm doing my budget as a soil-less project. I find it hard to let go and let the coning guide me. One thing I have learned is that I am given information and that it is what it is — information. I may have interpreted it wrong. Time will tell. I also have to use my own brain and be responsible for myself in making decisions. In the *Garden Workbook* it seems easier dealing with just plants. Can you offer any suggestions?

Machaelle's Answer: You need to think about the overall concept of the i/e partnership (involution/evolution partnership). I have stated and written numerous times that our job as humans is to remain focused on the evolutionary side of the equation. People don't believe that's enough, and that's partly why they like to tell nature how to do something. They underestimate the importance of that evolutionary dynamic they're involved in. If you remain entirely focused for the rest of your life and for the rest of all time solely on the evolutionary part of this equation, you will have fulfilled your life's purpose.

The equation seems easier in the Garden Workbook *because I'm providing you with the precise questions you need to ask to fulfill your role in the partnership with nature. There's less thinking required. Because SLGs can be so different and cover a wide range of projects, I can't provide long lists of questions. And that means your role is going to require more thought, more learning, more challenge, more chocolate and, consequently, more opportunity for personal evolution.*

APPENDICES

Appendix A

Kinesiology: The Tool for SLG Testing

KINESIOLOGY IS ANOTHER NAME for muscle testing. For those of you who use this method for getting information from nature or for testing the essences, you already have the tool in place for working with soil-less gardens.

For those of you who have never heard of such a thing but would like to work with soil-less gardens, just read on.

Kinesiology is simple. Anybody can do it because it uses your electrical system and your muscles. If you are alive, you have these two things. I know that sounds smart-mouthed of me, but I've learned that sometimes people refuse to believe that anything can be so simple. So they create a mental block — only "sensitive types" can do this, or only women can do this. It's not true. Kinesiology happens to be one of those simple things in life just waiting around to be learned and used by everyone.

I don't mean to intimidate you, but small children can learn to do kinesiology in about five minutes. It's mainly because it never occurred to them that they couldn't do it. If I tell them they have an electrical system, they don't argue with me about it — they just get on with the business of learning how to do simple testing. Actually, I *do* mean to intimidate you. Your first big hurdle will be whether or not you believe you have a viable electrical system that is capable of being tested. Here's a good test: Place a hand mirror under your nose. If you see breath marks, you have a strong electrical system. (If you don't see breath marks, call 911 — you're in trouble.) Now you can get on with learning how to use kinesiology!

If you've ever been to a chiropractor or holistic physician experienced in muscle testing, you've experienced kinesiology. The doctor tells you to stick out your arm and resist his pressure. It feels as if he is trying to push your arm down after he has told you not to let him do it. Everything is going fine, and then all of a sudden he presses and your arm flops down like an old fish. He is using kinesiology.

Simply stated, the body has within it and surrounding it an electrical network or grid. If a negative energy (that is, any physical object or energy vibration that does not maintain or enhance health and balance) is introduced into a person's energy field, his muscles, when having physical pressure applied, are unable to hold their strength. (The ability to hold muscle power is directly linked to the balance of the electrical system.) In other words, if pressure is applied to an individual's extended arm while his field is affected by a negative (energy), the arm will not be able to resist the pressure. It will weaken and fall to his side. If pressure is applied while affected by a positive (energy), the person will easily resist and the arm will hold its position. With SLGs, your coning team answers your yes/no questions by projecting a positive energy (yes) or negative energy (no) — whichever is appropriate — onto your electrical system. The "yes" or "no" you receive in the testing is actually from the team's projected answer. It is not an answer that has been concocted by you.

When a negative is placed in a person's energy field, his electrical system will immediately respond by short-circuiting or overloading. This makes it difficult for the muscles to maintain their strength and hold the position when any pressure is added. When a positive is placed within a person's energy field, the electrical system holds, and the muscles maintain their strength when pressure is applied.

This electrical/muscular relationship is a natural part of the human system. It is not mystical or magical. Kinesiology is the established method for reading that balance at any given moment.

If you have ever experienced muscle testing, you most likely participated in the above-described, two-person operation. You provided the extended arm, and the other person provided the pressure. Although efficient, this can be cumbersome when you want to test something on your own. Arm pumpers have the habit of disappearing when you need them. So for soil-less garden work you will be learning to self-test — no arm pumpers needed.

SELF-TESTING STEPS

1. THE CIRCUIT FINGERS. If you are right-handed: Place your left hand palm up. Connect the tip of your left thumb with the tip of the left little finger. (Not your index finger. I'm talking about your thumb and little finger.) If you are left-handed: Place your right hand palm up. Connect the tip of your right thumb with the tip of your right little finger. By connecting your thumb and little finger, you have just closed a major electrical circuit in your hand, and it is this circuit you will use for testing.

Fig. A: Circuit fingers–tip to tip

Before going on, look at the position you have just formed with your hand. If your thumb is touching the tip of your index or finger #1, laugh at yourself for not being able to follow directions, and change the position so you touch the tip of the thumb with the tip of the little finger. Most likely this will not feel at all comfortable to you. That is because you normally don't put your fingers in this position and they might feel a little stiff. If you are feeling awkwardness, you've got the first step of the test position! In time, the hand and fingers will adjust to being put in this position and it will feel fine.

Fig. B: Circuit fingers–pad to pad

Circuit fingers can touch tip to tip *(Fig. A),* finger pad to finger pad *(Fig. B),* or thumb resting on top of the little finger's nail *(Fig. C).* I rest my thumb on top of my little finger. And I suggest this position for anyone with long nails. You're not required to impale yourselves for this.

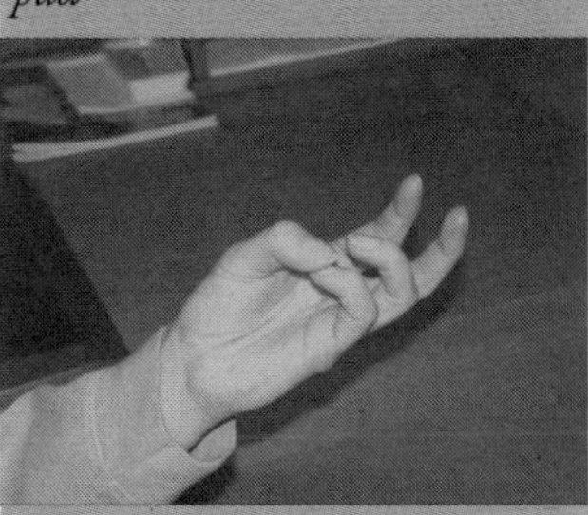

Fig. C: Circuit fingers–thumb on top of little finger

When you have the circuit fingers in position, they form a circle. If you straighten fingers 1, 2 and 3 a bit, you'll get them out of the way and you'll see the circle.

2. THE TEST FINGERS AND TESTING POSITION. To test the circuit (the means by which you will apply pressure), place the test fingers, thumb and index finger of your other hand *(Fig. D),* inside the circle you have created by connecting your circuit thumb and little finger. The test fingers (thumb/index finger) should be right under the circuit fingers (thumb/little finger), touching them, with your test thumb resting against the underside of your circuit thumb and your test index finger resting against the underside of your circuit little finger. (See *Fig. A,* next page.)

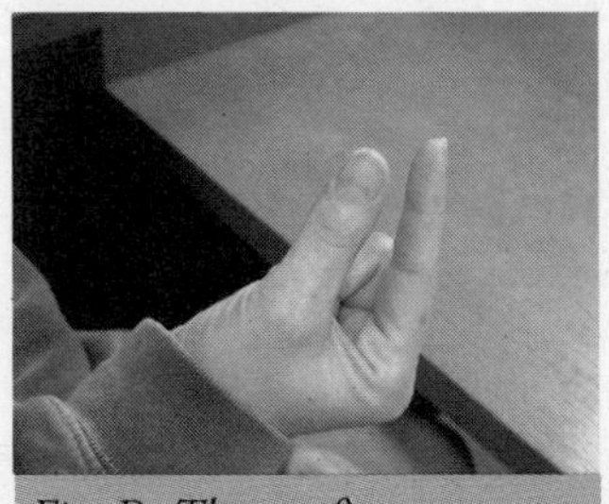

Fig. D: The test fingers

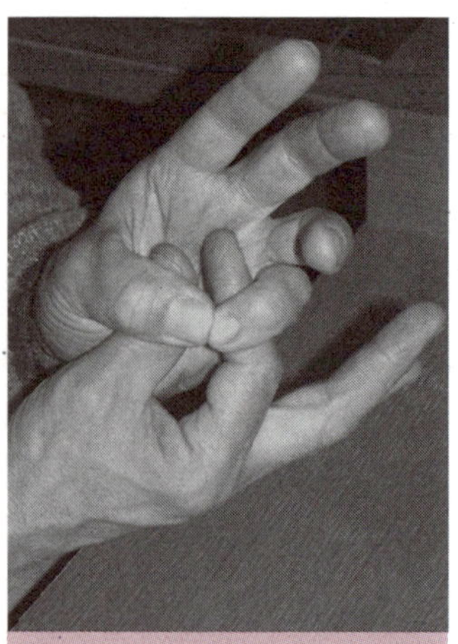

Fig. A: The testing position

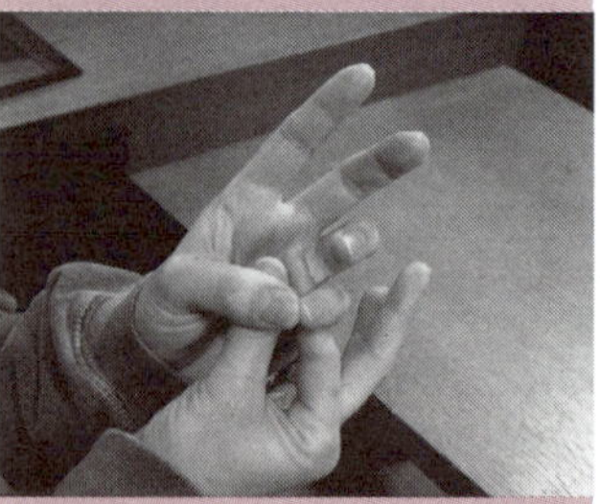

Fig. B: Positive response with the circuit fingers still closed

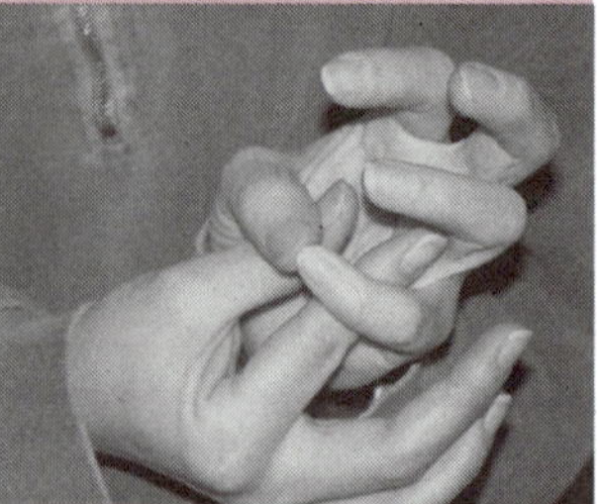

Fig. C: Another view of the positive response

Don't try to make a circle with your test fingers. They are just placed inside the circuit fingers that do form a circle. It will look like you have two "sticks" inserted inside a circle.

3. POSITIVE RESPONSE. Keeping this position, ask yourself a simple question in which you already know the answer to be "yes." ("Is my name _____?") Once you've asked the question, press your circuit fingers together, keeping them in the circular position. *Using the same amount of pressure,* try to press apart or separate the circuit fingers with your test fingers. Press the lower thumb against the upper thumb, and the lower index finger against the upper little finger. The action of your test fingers will look like scissors separating as you apply pressure to your circuit fingers. Your testing fingers, the fingers inserted in the circuit circle, will remain in position within the circle. *(Figs. B and C)* All you are doing is using these two testing fingers to apply pressure to the outer two circuit fingers. Don't try to pull your test fingers vertically up through your circuit fingers.

The circuit position described in step 1 corresponds to the position you take when you stick your arm out for the physician. The testing position in step 2 is in place of the physician or other convenient arm pumper. After you ask the yes/no question and you press your circuit fingers tip-to-tip, that is equal to the doctor saying, "Resist my pressure." Your circuit fingers now correspond to your outstretched, stiffened arm. Trying to push apart those fingers with your testing fingers is equal to the doctor pressing down on your arm.

If the answer to the question is positive (if your name is what you think it is!), you will not be able to easily push apart the circuit fingers. The electrical circuit will hold, your muscles will maintain their strength, and your circuit fingers will not separate. You will feel the strength in that circuit.

CALIBRATING THE FINGER PRESSURE: Be sure the amount of pressure holding the circuit fingers together is equal to the amount of your testing fingers pressing against them. Also, do not use a pumping action (pressing against your circuit fingers several times in rapid succession) when applying pressure to your circuit fingers. Use an equal and continuous pressure.

Play with this a bit. Ask a few more yes/no questions that have positive answers. Now, I know it is going to seem that if you already know the answer to be "yes," you are probably "throwing" the test. Well, you are. This is your tool for calibrating your fingers for feeling the strong positive. You are asking yourself a question that has a positive answer. If your circuit fingers are separating, you are applying too much pressure with your testing fingers. Or you are not putting enough pressure into holding your circuit fingers together. You need to keep asking the question and play with the testing until you feel pressure in all four fingers and the pressure in your testing fingers is not separating your circuit fingers. You don't have to break or strain your fingers for this; just use enough pressure to make them feel alive, connected and alert. When this happens, now you have a clear positive kinesiology response.

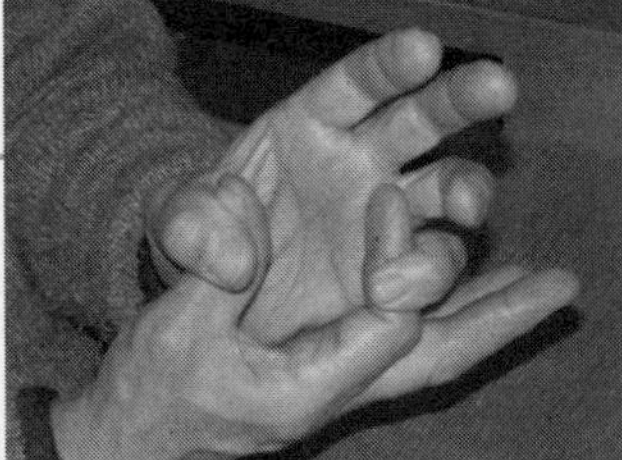

Fig. D: Negative response – a lot of separation

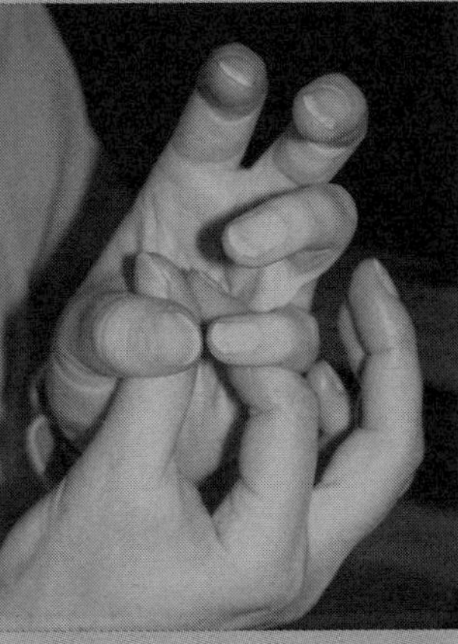

Fig. E: Negative response – a little separation

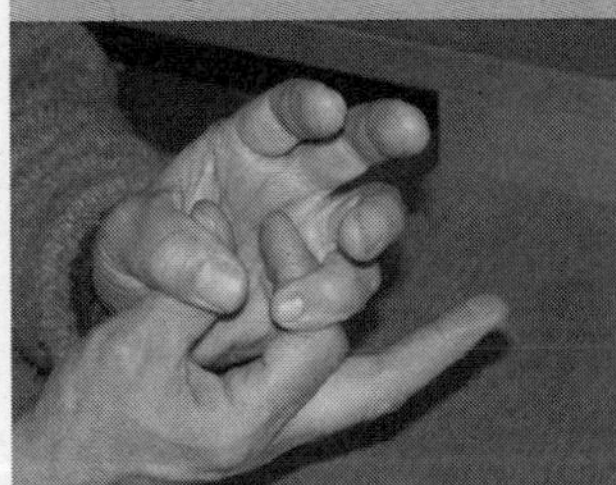

Fig. F: Negative response – medium separation

4. NEGATIVE RESPONSE. Once you have a clear sense of the positive response, ask yourself a question that has a negative answer. Again press your circuit fingers together and, using equal pressure, press against the circuit fingers with the test fingers. This time, if the testing-fingers' pressure is equal to the circuit-fingers' pressure, the electrical circuit will break, and the circuit fingers will weaken and separate. Because the electrical circuit is broken, the muscles in the circuit fingers do not have the power to hold the fingers together. In a positive state the electrical circuit holds, and the muscles have the power to keep the two fingers together.

DIFFERENT STYLES IN HOW THE FINGERS SEPARATE: How much your circuit fingers separate depends on your personal style. Some people's fingers separate a lot. *(Fig. D)* Other's barely separate at all. *(Fig. E)* Mine separate about a quarter of an inch. *(Fig. F)* Some people's fingers won't separate at all, but they'll definitely feel the fingers weaken when pressure is applied during a "no" answer. Let your personal style develop naturally.

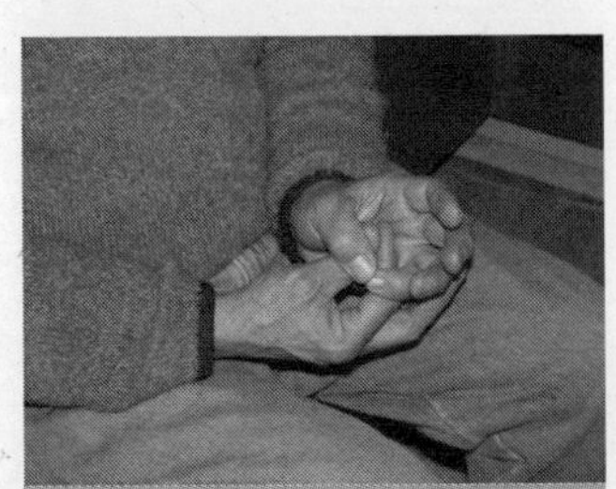

Fig. G: Forearms resting in a person's lap while testing

RESTING YOUR FOREARMS: If you are having a little trouble feeling anything, do your testing with your forearms resting in your lap. *(Fig. G)* This way you won't be using your muscles to hold your arms up while you are trying to test.

CALIBRATING AND EQUALIZING THE PRESSURE used by the circuit fingers and the testing fingers for negative responses: Play with negative questions and continue adjusting the pressure between your circuit and test fingers until you get a clear negative response.

When you're feeling a solid separation, return to positive questions. Once again, get a good feeling for the strength between your circuit fingers when the electricity is in a positive state. Then ask a negative question and feel the weakness when the electricity is in a negative state. Practice your testing by alternating the questions.

In the beginning, you may feel only a slight difference between the two. With practice, that difference will become more pronounced. For now, it is just a matter of trusting what you have learned — and practicing.

THE TESTING CALIBRATION: Especially in the beginning, and even sometimes after you have been doing kinesiology successfully for awhile, you may lose the strong feeling of the positive response and the weakness of the negative. You've just lost the equal pressure between your circuit and testing fingers and one set is overpowering the other.

When this happens, just back away from whatever you are trying to test and do a testing calibration. Ask yourself a question that you know has a positive answer and test for the response. Adjust the pressure between your testing and circuit fingers until you feel a strong, positive response. Play with this a bit and get a good feel for the strength of the positive responses.

Then switch to questions that have a negative response and play around with the pressure until you feel a clear breaking of the circuit.

After this, alternate your questions between positive and negative a few times and test the answer. In no time, you'll have the "kinesiology feel" back and you can resume the work you were testing where you left off.

Don't forget the overall concept behind kinesiology. What enhances our body, mind and soul makes us strong. Together, our body, mind and soul create a holistic environment that, when balanced, is strong and solid. If something enters that environment and negates or challenges the balance, the environment is weakened. That strength or weakness registers in the electrical system, and it can be discerned through a muscle-testing technique — kinesiology.

IMPORTANT: Do a testing calibration as soon as your testing feels a little off or funny to you so that you won't waste time doing a bunch of testing and not being sure of the accuracy of those answers. If you want accurate results in anything you are testing and especially when working with soil-less gardens, you will have to retest those parts where you felt your testing to be suspect. So do a testing calibration sooner rather than later. It'll save time in the long run.

KINESIOLOGY TIPS

THE NEED TO SWITCH HANDS: If you are having trouble feeling the electrical circuit in the circuit fingers, try switching hands — the circuit fingers become the test fingers and vice versa. Most people who are right-handed have this particular electrical circuitry in their left hand. Left-handers generally have the circuitry in their right hand. But sometimes a right-hander has the circuitry in the right hand and a left-hander has it in the left hand. Or some heavy-handed teacher along the way forbid you to write with the left hand and required you to learn to write with your right hand. You may be one of those people. If you are ambidextrous, choose the circuit hand that gives you the clearest responses. Before deciding which to use, give yourself a couple of weeks of testing using one hand as the circuit hand to get a good feel for its responses before trying the other hand.

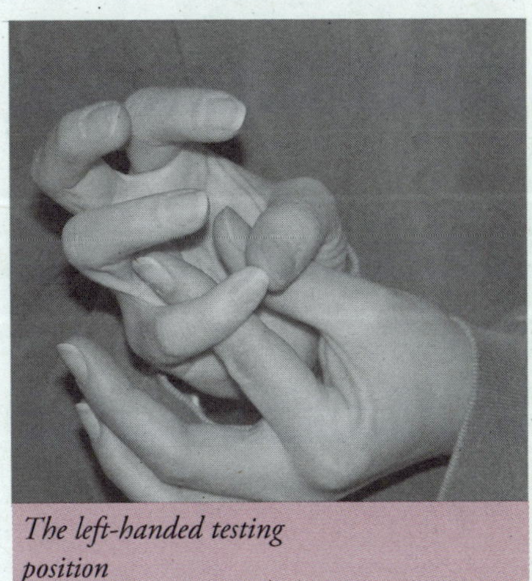

The left-handed testing position

ALTERNATIVE KINESIOLOGY TESTING METHOD

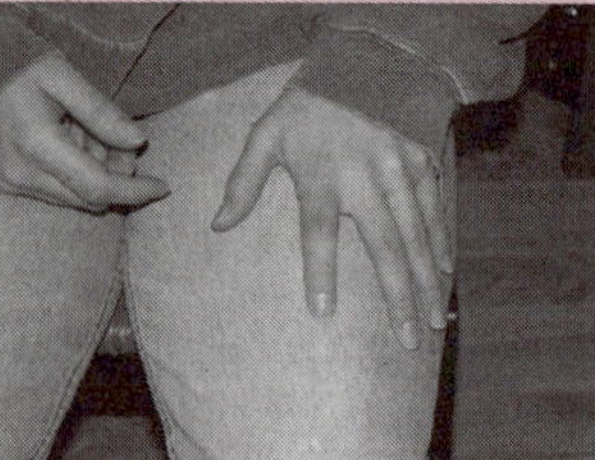

Fig. A: Connecting the finger-to-thigh circuit

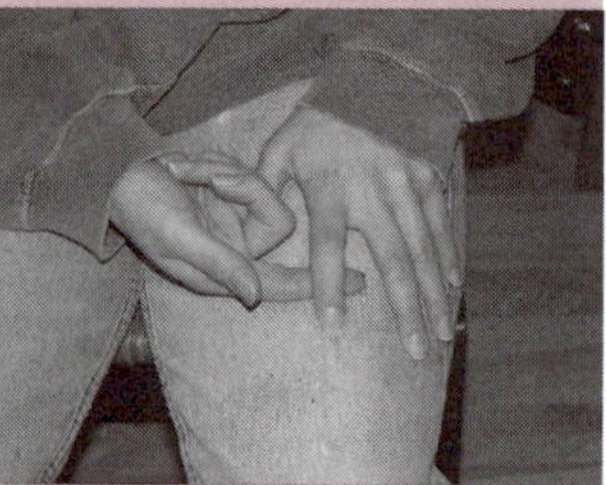

Fig. B: The testing position

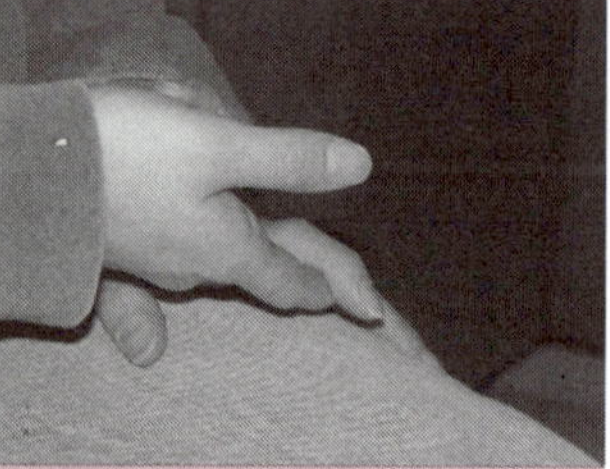

Fig. C: Negative response–the circuit breaks and the test finger lifts

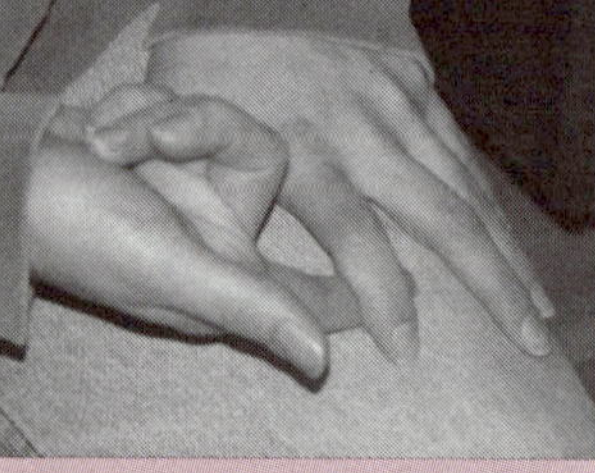

Fig. D: Positive response–the circuit remains strong and the test finger remains connected

KINESIOLOGY TESTING AND INJURIES: If you have an injury such as a muscle sprain in either hand or arm, don't try to learn kinesiology until you have healed. Kinesiology is muscle testing, and a muscle injury will interfere with the testing — and the testing will interfere with the healing of the muscle injury.

AN ALTERNATIVE KINESIOLOGY TESTING METHOD: There is a method of kinesiology that uses the pointing finger of your test hand and your leg. If you have a physical impairment, this may be easier for you to use.

- Connecting the circuit: Place the pointing finger (index finger) of your test hand on top of the center of your thigh. (Left test hand on left thigh or right test hand on right thigh.) This finger should lay flat on the leg. Your other fingers may be in whatever position is comfortable. *(Fig. A)*
- The testing position: Place the pointing finger of the other hand in a face-up position under the first knuckle of the test finger. *(Fig. B)* Make sure you are using the knuckle section, not just the tip of the finger (the first knuckle of each pointing finger should be in contact).
- The electrical circuit that you are using runs down the center of the leg and you are connecting it with the circuit in your finger.
- To test: ask a question, press your test finger down on the leg, then, using equal pressure, try to lift up the finger of the test hand.
- If the circuit breaks easily and the testing finger lifts, the answer is negative. *(Fig. C)*
- If the circuit holds and it is difficult to lift the finger, the answer is positive. *(Fig. D)*

LEARNING KINESIOLOGY IN THE RIGHT ENVIRONMENT: When first learning kinesiology, do yourself a favor and set aside some quiet time to go through the instructions and play with the testing. Trying to learn this while riding the New York subway during evening rush hour isn't going to give you the break you need. But once you have learned it, you will be able to test all kinds of things while riding the subway.

THE KINESIOLOGY BODY/MIND CONNECTION: Sometimes I meet people who are trying to learn kinesiology and aren't having much luck. They've gotten frustrated, decided this isn't for them, and have gone on to try to learn another means of testing. Well, I'll listen to them explain what they did, and before they know it, I've verbally tricked them with a couple of suggestions about their testing, which they try, and they begin feeling kinesiology for the first time — a strong "yes" and a clear "no." The problem wasn't kinesiology. Everyone, as I have said, has an electrical system. The problem was that they wanted to learn it so much that they became overly anxious and tense — they blocked.

So, since you won't have me around to trick you, if you suspect you're blocking, go on to something else. Then trick yourself. When you care the least about whether or not you learn kinesiology, start playing with it again. Approach it as if it were a game. Then you'll feel the strength and weakness in the fingers.

TROUBLESHOOTING YOUR TESTING: Now, suppose the testing has been working fine, and then suddenly you can't get a clear result (what I call a "definite maybe") or you get no result at all. Check the following:

1. *SLOPPY TESTING.* You try to press apart the fingers before applying pressure between the circuit fingers. This happens especially when we've been testing for awhile and become over-confident or do the testing very quickly. I think it happens to all of us from time to time and serves to remind us to keep our attention on the matter at hand. (Excuse the pun.)

Especially in the beginning, start a kinesiology session by first feeling a few positive and negative responses. Ask yourself some of those obvious questions. Or simply say several times, "Let me feel a positive." (Test.) "Let me feel a negative." (Test.) This will serve as a kind of warm-up and remind you what positive and negative feel like before you start. If the testing begins to feel sloppy in the middle of some soil-less garden work, stop the testing and do a kinesiology calibration (p. 114) to get the pressure equalized on both hands again.

2. *External distractions.* Trying to test in a noisy or active area can cause you to lose concentration. The testing will feel unsure or contradict itself if you double-check the results. Often, simply moving to a quiet, calm spot and concentrating on what you are doing will be just what's needed for successful testing.

3. *Focus or concentration.* Even in a quiet spot, one's mind may wander and the testing will feel fuzzy, weak or contradictory. It is important to concentrate throughout the process. Check how you are feeling. If you're tired, I suggest you not try to test until you have rested a bit. And if you have to go to the bathroom, do it. That little situation is a sure concentration-destroyer.

4. *The question isn't clear.* A key to kinesiology is asking a simple yes/no question, not two questions in one, each having a possible yes/no answer. If your testing isn't working, first check your hand positions. Next, review your question, and make sure you are asking only one question. And, while you're asking a question, don't think ahead to the next question! Your fingers won't know which to answer.

5. *Match your intent with your wording.* If you are prone to saying, "Oh, I didn't mean to say that!" when you talk to others, this might be an area you need to work on.

A woman at one of our workshops asked me about some strange answers she had gotten about what to feed her cat. She had asked, "What kinds of food would make my cat happy?" She got weird answers like chocolate, catnip, steak . . . I pointed out that she probably asked the wrong question. She meant to ask what foods would make her cat healthy. She was a little surprised. She thought that this was the question she had originally asked. In short, her question and her intent didn't match.

6. *You must want to accept the results.* If you enter a kinesiology test not wanting to "hear" the answer, for whatever reason, you can override the test with your emotions and your will. This is true for conventional situations as well. If you really don't want something to work for you, it won't work. That's our personal power dictating the outcome.

7. *Testing situations that are personal emotional minefields.* If you are trying to do testing during a situation that is especially emotional for you, that deeply stirs your emotions, or if you are trying to ask a question in which you have a strong, personal investment in the answer — such as, "Should I buy this beautiful $850,000 house?"— I suggest that you not test until you are calmer or get some emotional distance from the situation. During such times, you are walking a very fine line between a clear test and a test that your desires are overriding. Kinesiology as a tool is not the issue here. It is the condition or intent of the tester. In fact, some questions just shouldn't be asked, but which questions shouldn't be asked is relative to who is doing the asking. We each need to develop discernment around which questions are appropriate for us to ask.

When I am involved with testing during emotionally stressful times, I stop for a moment, collect my thoughts and make a commitment to concentrate on the testing only. If I must test an emotionally charged question or a question about something I have a personal investment in, I stop a moment, commit myself to the test and open myself to receiving the answer and not the answer I might desire.

> IMPORTANT: I suggest that you make it a rule not to ask questions like "Do I have cancer?" Your ability to perceive the answer either through kinesiology or intuition is questionable. It is too emotionally charged. If you have a suspicion, get a solid diagnosis from the allopathic medical establishment, and make sure you get two or three "second opinions."

8. *A sudden breakdown of your testing.* If your testing has been going along just fine and you suddenly begin to get contradictory or "mushy" test results, consider that this may not be a good day for you to do this particular work.

- You could be too tired to hold your focus on the testing.
- *Or you may need to drink water!* If you are dehydrated, your electrical system will feel weak during kinesiology testing. It's like you need to put water in your battery.

- Or you may need to quickly rebalance your body's electrical system. To do this, take one dose (10–12 drops) of ETS Plus for Humans. That may be just what you need for clear kinesiology results.

If your testing "breaks down": If your kinesiology is "breaking down" and your testing results seem odd (you're getting all negative responses or all positive responses), you need to back away from the testing and take a short break. You are nervous, distracted, in need of water and/or tired.

- If you are tired, take a short break, eat some protein and return to the testing where you left off or start the testing again from the top if you're not sure about the accuracy of your previous testing.
- If you are distracted, take a short break, do what you need to do to eliminate the distraction and return to the testing where you left off.
- If you are nervous, take a short break, regroup, give yourself a pep talk and relax. Then return to the testing where you left off. If you need water, drink.

Calibrating your testing: Before resuming the testing, "calibrate your fingers" and get the feel for your kinesiology yes/no responses.

- Do this by first asking yourself a question you already know the answer to and is guaranteed to give you a positive response, such as "Is my name ________?" Now test.
- If you don't get the positive response, your fingers need calibrating. The pressure you are using on both hands isn't equal.
- Adjust the pressure (back off the pressure you are exerting to hold the circuit fingers together or the pressure you're exerting with the two testing fingers) until you get a clear, strong positive response.
- Then ask the question and continue adjusting the pressure when needed.
- Do this until you feel you're getting a consistent positive response to your question.

Now repeat this process, only this time ask a question that you know will give you a negative response. "Do I weigh 1863 pounds?" Or "Is my name Donald Duck?" (If your name is actually Donald Duck, you'll need to pick another name, like "Minnie Mouse.") Continue with the kinesiology calibration process until you feel you're getting a consistent negative response to your question. Then resume the SLG testing.

IF YOU'RE HAVING TROUBLE LEARNING KINESIOLOGY FROM A BOOK AND YOU WANT MORE HELP: You're in luck. I give an excellent demonstration (if I don't mind saying so myself) of how to do kinesiology with the DVD/video *Working with Nature in Soil-less Gardens.*

Appendix B

Perelandra SLG Calibration Process Steps for Groups or Teams

THIS IS A PROCESS THAT can be used to work with nature to:

1. align an individual or team with a project's direction, definition and purpose,

2. calibrate to add a new member to a project's team and align the new team with the project's DDP,

3. calibrate to remove a member from the team, re-form the new team and align this new team with the project's DDP, and

4. realign an individual or team with the project's DDP when needed during times of development or trouble.

INSTRUCTIONS FOR DOING THE PROCESS WITH GROUPS OR TEAMS

Before beginning, briefly discuss the general purpose of the calibration. Be sure you have a watch or clock and everyone else has pen and paper.

1. State aloud:

> I would like to open a Calibration Process for this project. I would like to connect with (open the 4-point coning for your SLG project).

You must include the higher self of each person attending the calibration. Addressing each person one by one, request that this person's higher self be included in the coning. Wait 10 seconds for each connection to occur. Have everyone take *one dose* (10–12 drops) of ETS Plus for Humans.

2. State aloud:

I request that Pan shift forward in the coning.

This is needed since Pan is the coning partner that actually does the balancing work during the Calibration Process.

3. Review the intent and steps that you will move through. When there are no further questions, continue the Calibration Process.

4. State aloud:

This is a calibration for the following purpose (insert purpose).

5. Ask Pan:

How much time is needed for this calibration?

Do a sequential kinesiology test to determine how many minutes are needed. This can be 5 minutes to 30 minutes. Tell the group the amount of time needed to complete the calibration. Remind them to allow the mind to relax during this period and remain comfortably focused on the intent. Everyone is to remain quiet for the designated time period.

6. Restate aloud:

We would like to begin the calibration for (insert purpose) now.

7. Calibration time completed: Gently announce the end of the time period. Allow a minute or so for the individuals to shift their focus back to the room. Have everyone take *one dose* (10–12 drops) of ETS Plus for Humans.
Then balance and stabilize this SLG Calibration Process. (See p. 62 for the steps.)

8. Discuss any insights, pictures or feelings that were experienced during the calibration.

9. Close the coning. Remember to disconnect each person's higher self from the coning. Do this by addressing each person one by one and requesting that their higher self be disconnected from the coning. Wait 5 seconds between disconnection. Have everyone take *one dose* (10–12 drops) of ETS Plus for Humans.

To Add or Remove a Member of the Project's Team

CALIBRATION PROCESS STEPS FOR TEAMS

When using this process to add a new member to a project's team or remove a member from the team, the calibration is done in two phases.

MAKE THE FOLLOWING CHANGES IN THE PROCESS FOR STEP 4:*

4A. State the intent:

> This is a calibration of the project team to the project's definition, direction and purpose.

4B. PHASE 1: State aloud,

> In the first phase, we would like a calibration of the project's team to add/remove (insert name of person) to/from the team.

4C. Kinesiology test to determine how many minutes are needed.

4D. Restate aloud:

> We would like to begin the calibration of the project's team to add/remove (insert name of person) to/from the team now.

When complete, shift the focus to phase 2 of the calibration.

4E. PHASE 2: State aloud,

> The focus in this phase is the calibration of the new team to the project's present definition, direction and purpose.

4F. Kinesiology test to determine how many minutes are needed.

4G. Restate aloud:

> We would like to begin the calibration of the new team to the project's present definition, direction and purpose, now.

RESUME THE CALIBRATION PROCESS, STEPS 7 THROUGH 9 ON P. 124.

Suggestions for the Leader

- You will be leading the process so you will experience it differently than if you were participating. You will be focused on moving the group smoothly through the process with clarity and precision.

* Steps 4A through 4G replace step 4 on p. 124. Skip steps 5 and 6 (p. 124). Then do steps 7 through 9 to complete the Calibration Process for adding or removing a member of the team.

- You must understand the process and the intent behind it. You will be holding the intent and leading the process through its stages. Be precise and clear with all statements. Review the process ahead of time and be prepared to lead the group.

- It is your responsibility to maintain the focus for the group. If something happens during the process to diffuse the focus (someone sneezes, a door slams, a phone rings), gently bring the group back to the focus of the calibration. Restate the calibration intent and proceed with the calibration for the time remaining.

- Be sure that everyone participating in the process has read, discussed and understands the Calibration Process steps. If there is a new member of the group, take time before the day of the calibration to explain the process to him and answer any questions he might have. Take time again before beginning the process to allow anyone in the group to ask questions or ask for further clarification. (For important background information on the Calibration Process and how nature works with us in a project calibration, see DVD/Video 3: *Working with Nature in Soil-less Gardens.*)

Suggestions for the Participants

- Keep the focus on your working relationship with nature. Don't make this a spiritual experience.

- There needs to be only one intent for each project calibration. Be sure you don't silently add things to the intent or change the intent in any way. It is important that each person in the group focus only on the stated intent.

- Everyone needs to be comfortably focused on what's happening and relaxed. Don't over-do (try too hard) or let your mind wander to other things. Eliminate any possible physical distractions before the process begins: turn off radios, turn off phone ringers, sequester pets, etc.

- If you have any questions about the process, if anything in the process or in the stated intent is not clear to you, say something right away. If at any point in the process, you are not ready to focus or shift a focus, let the person leading the calibration know.

BLANK CHARTS

Perelandra SLG Troubleshooting Chart

Project: ______________________ Date: __________

1. Deva of project, Pan, the appropriate WB connection and your h.s. (Take one dose ETS Plus for Humans. Wait 10 seconds.)
2. Identify & describe issue/problem: ______________________
3. State intent: "I request that the coning focus be directed to the above stated issue/problem."
4. Ask: "Is a Troubleshooting Process needed for this issue/problem?" ☐ yes ☐ no (If "no," go to Step 9.)

5. SLG TROUBLESHOOTING LISTS

___ **Environmental Troubleshooting List**

- ___ New/Necessary Equipment/Supplies: __________
- ___ Organize/Reorganize/Relocate: __________
- ___ Clean: __________
- ___ Repair/Upgrade: __________
- ___ Staff/Consultant change: __________
- ___ Meeting/Brainstorming/Funding/Resources: __________
- ___ Additional needs: __________

___ **Energy Process Troubleshooting List**

___ ETS Plus for Soil-less Gardens Process: Do a General Balancing. ☐ Test/treat as unit ☐ Test/treat separately ☐ Telegraph Test

# of drops for treating as a unit: __________

___ Order / # of drops: __________

___ Organization / # of drops: __________

___ Life Vitality / # of drops: __________

___ ETS Plus for Soil Process

# of drops: __________

___ Energy Cleansing Process ☐ Project Framework ☐ Mental-Level Activity ☐ Both

Bal / # of drops: __________

Stab / # of drops: __________

___ Battle Energy Release Process

Bal / # of drops: __________

Stab / # of drops: __________

___ Balancing and Stabilizing Process

___ Full project/general balancing

Bal / # of drops: __________

Stab / # of drops: __________

___ Specific problem/issue stated in Step 2

Bal / # of drops: __________

Stab / # of drops: __________

___ Essences Process (NS application)

___ General Solution

Sol/Dosage: __________

___ Solution for specific problem

Sol/Dosage: __________

___ Triangulation Process	1) ☐ point	☐ link	2) ☐ point	☐ link	3) ☐ point	☐ link
Bal / # of drops:						
Stab / # of drops:						

___ SLG Calibration Process ☐ Align to DDP ☐ Add to team ☐ Remove from team ☐ Realign to DDP ☐ Align project/issue to DDP

Bal / # of drops: __________

Stab / # of drops: __________

(Troubleshooting Process Steps continued on back)

SLG Troubleshooting Process Options

___ **Human Troubleshooting List** (Only test the processes you use.)

Include #1 & 2 when testing for the Step 6 order for the SLG Troubleshooting.

___ 1. ETS Plus for Humans

___ 2. Essences Telegraph Test

Sol / Dosage: ___

See the *SLG Companion,* pp. 67–68 for determining the order of the remaining Options.

___ 3. Personal Calibration Process

___ 4. MAP

___ 5. MAP/Calibration

___ ***Workbook II*** **Troubleshooting Process & Chart**

___ **Microbial Balancing Program Troubleshooting Process & Chart**

(Continuation of the Troubleshooting Process Steps)

6. Order the processes to be done in the Energy Process List, plus the Human List #1 and @2. (Place order to left of list.)

7. Take one dose of ETS Plus for Humans. (Take this dose for any needed personal balancing resulting from this troubleshooting work.)

8. Troubleshooting recheck date: ☐ yes ☐ clear Date: ___

9. Close coning. Take one dose ETS Plus for Humans.

Perelandra SLG Troubleshooting Chart

Project: ______________________________ Date: ____________

1. Deva of project, Pan, the appropriate WB connection and your h.s. (Take one dose ETS Plus for Humans. Wait 10 seconds.)
2. Identify & describe issue/problem: ______________________________
3. State intent: "I request that the coning focus be directed to the above stated issue/problem."
4. Ask: "Is a Troubleshooting Process needed for this issue/problem?" ☐ yes ☐ no (If "no," go to Step 9.)

5. SLG Troubleshooting Lists

___ **Environmental Troubleshooting List**

- ___ New/Necessary Equipment/Supplies: ______________
- ___ Organize/Reorganize/Relocate: ______________
- ___ Clean: ______________
- ___ Repair/Upgrade: ______________
- ___ Staff/Consultant change: ______________
- ___ Meeting/Brainstorming/Funding/Resources: ______________
- ___ Additional needs: ______________

___ **Energy Process Troubleshooting List**

___ ETS Plus for Soil-less Gardens Process: Do a General Balancing. ☐ Test/treat as unit ☐ Test/treat separately ☐ Telegraph Test

# of drops for treating as a unit: ______________

- ___ Order / # of drops: ______________
- ___ Organization / # of drops: ______________
- ___ Life Vitality / # of drops: ______________

___ ETS Plus for Soil Process

# of drops: ______________

___ Energy Cleansing Process ☐ Project Framework ☐ Mental-Level Activity ☐ Both

Bal / # of drops: ______________

Stab / # of drops: ______________

___ Battle Energy Release Process

Bal / # of drops: ______________

Stab / # of drops: ______________

___ Balancing and Stabilizing Process

___ Full project/general balancing

Bal / # of drops: ______________

Stab / # of drops: ______________

___ Specific problem/issue stated in Step 2

Bal / # of drops: ______________

Stab / # of drops: ______________

___ Essences Process (NS application)

___ General Solution

Sol/Dosage: ______________

___ Solution for specific problem

Sol/Dosage: ______________

___ Triangulation Process	1) ☐ point	☐ link	2) ☐ point	☐ link	3) ☐ point	☐ link
Bal / # of drops:						
Stab / # of drops:						

___ SLG Calibration Process ☐ Align to DDP ☐ Add to team ☐ Remove from team ☐ Realign to DDP ☐ Align project/issue to DDP

Bal / # of drops: ______________

Stab / # of drops: ______________

(Troubleshooting Process Steps continued on back)

SLG Troubleshooting Process Options

___ **Human Troubleshooting List** (Only test the processes you use.)

Include #1 & 2 when testing for the Step 6 order for the SLG Troubleshooting.

___ 1. ETS Plus for Humans
___ 2. Essences Telegraph Test
Sol / Dosage: ______________________

See the *SLG Companion,* pp. 67–68 for determining the order of the remaining Options.

___ 3. Personal Calibration Process
___ 4. MAP
___ 5. MAP/Calibration

___ ***Workbook II*** **Troubleshooting Process & Chart**

___ **Microbial Balancing Program Troubleshooting Process & Chart**

(Continuation of the Troubleshooting Process Steps)

6. Order the processes to be done in the Energy Process List, plus the Human List #1 and @2. (Place order to left of list.)

7. Take one dose of ETS Plus for Humans. (Take this dose for any needed personal balancing resulting from this troubleshooting work.)

8. Troubleshooting recheck date: ☐ yes ☐ clear Date: ______________________

9. Close coning. Take one dose ETS Plus for Humans.

Perelandra SLG Troubleshooting Chart

Project: ______________________________ Date: ______________

1. Deva of project, Pan, the appropriate WB connection and your h.s. (Take one dose ETS Plus for Humans. Wait 10 seconds.)
2. Identify & describe issue/problem: ______________________________

____

3. State intent: "I request that the coning focus be directed to the above stated issue/problem."
4. Ask: "Is a Troubleshooting Process needed for this issue/problem?" ☐ yes ☐ no (If "no," go to Step 9.)

5. SLG TROUBLESHOOTING LISTS

___ **Environmental Troubleshooting List**

- ___ New/Necessary Equipment/Supplies: ______________
- ___ Organize/Reorganize/Relocate: ______________
- ___ Clean: ______________
- ___ Repair/Upgrade: ______________
- ___ Staff/Consultant change: ______________
- ___ Meeting/Brainstorming/Funding/Resources: ______________
- ___ Additional needs: ______________

___ **Energy Process Troubleshooting List**

___ ETS Plus for Soil-less Gardens Process: Do a General Balancing. ☐ Test/treat as unit ☐ Test/treat separately ☐ Telegraph Test

# of drops for treating as a unit: ______________

___ Order / # of drops: ______________

___ Organization / # of drops: ______________

___ Life Vitality / # of drops: ______________

___ ETS Plus for Soil Process

# of drops: ______________

___ Energy Cleansing Process ☐ Project Framework ☐ Mental-Level Activity ☐ Both

Bal / # of drops: ______________

Stab / # of drops: ______________

___ Battle Energy Release Process

Bal / # of drops: ______________

Stab / # of drops: ______________

___ Balancing and Stabilizing Process

___ Full project/general balancing

Bal / # of drops: ______________

Stab / # of drops: ______________

___ Specific problem/issue stated in Step 2

Bal / # of drops: ______________

Stab / # of drops: ______________

___ Essences Process (NS application)

___ General Solution

Sol/Dosage: ______________

___ Solution for specific problem

Sol/Dosage: ______________

___ Triangulation Process

	1) ☐ point	☐ link	2) ☐ point	☐ link	3) ☐ point	☐ link
Bal / # of drops:						
Stab / # of drops:						

___ SLG Calibration Process ☐ Align to DDP ☐ Add to team ☐ Remove from team ☐ Realign to DDP ☐ Align project/issue to DDP

Bal / # of drops: ______________

Stab / # of drops: ______________

(Troubleshooting Process Steps continued on back)

SLG Troubleshooting Process Options

___ **Human Troubleshooting List** (Only test the processes you use.)

Include #1 & 2 when testing for the Step 6 order for the SLG Troubleshooting.

___ 1. ETS Plus for Humans
___ 2. Essences Telegraph Test
Sol / Dosage: ______

See the *SLG Companion,* pp. 67–68 for determining the order of the remaining Options.

___ 3. Personal Calibration Process
___ 4. MAP
___ 5. MAP/Calibration

___ ***Workbook II* Troubleshooting Process & Chart**

___ **Microbial Balancing Program Troubleshooting Process & Chart**

(Continuation of the Troubleshooting Process Steps)

6. Order the processes to be done in the Energy Process List, plus the Human List #1 and @2. (Place order to left of list.)

7. Take one dose of ETS Plus for Humans. (Take this dose for any needed personal balancing resulting from this troubleshooting work.)

8. Troubleshooting recheck date: ☐ yes ☐ clear Date: ______

9. Close coning. Take one dose ETS Plus for Humans.

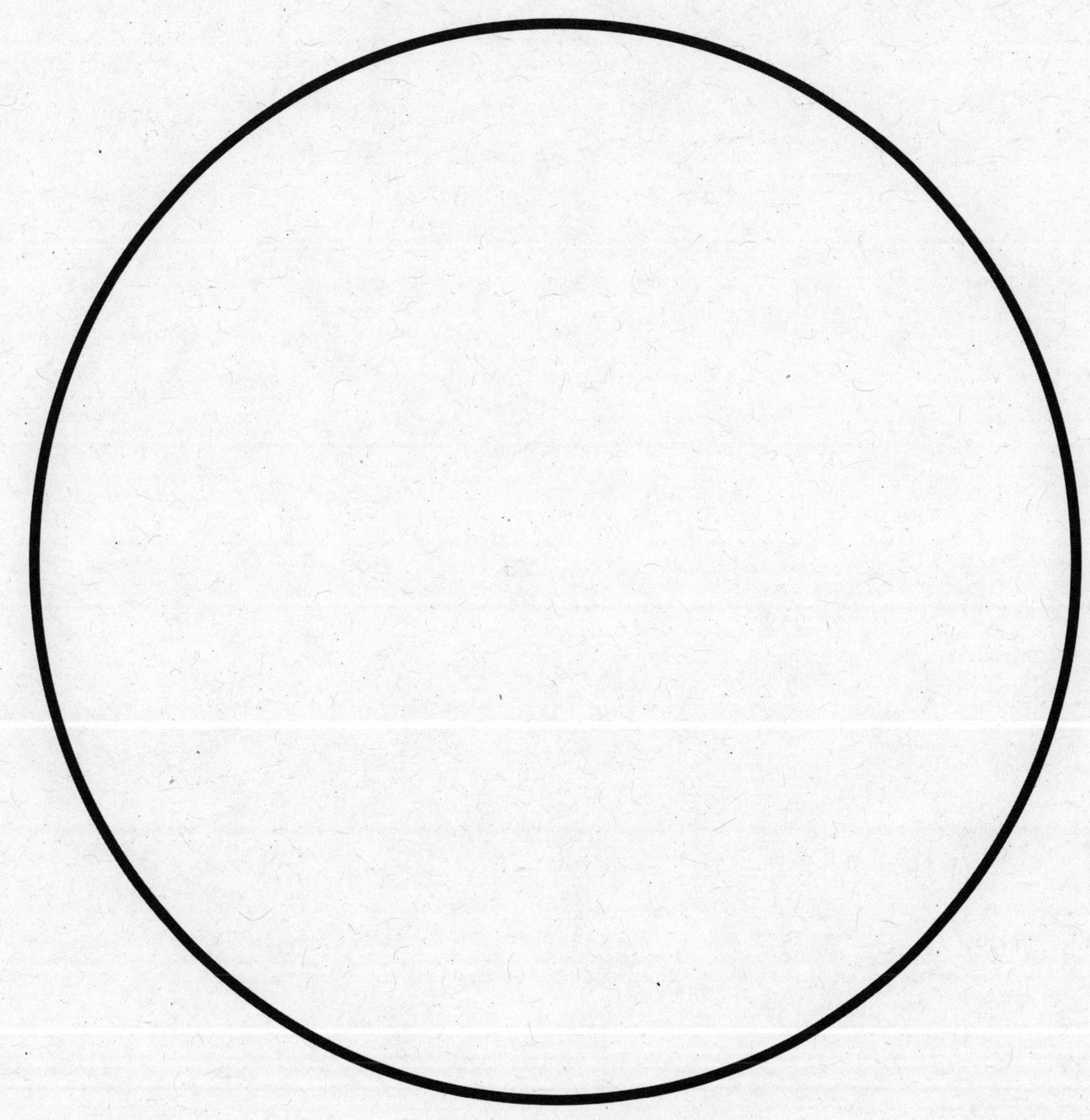

Circle Template for the Drawing Impaired.
Use with the SLG Energy Cleansing Process.